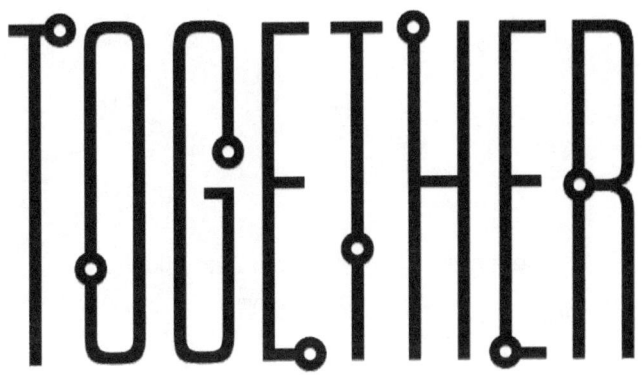

AI AND HUMAN.
ON THE SAME SIDE.

ZOLTAN ANDREJKOVICS

#ARTIFICIALINTELLIGENCE #AI #ROBOTICS
#FUTURISM #ARTIFICIALSUPERINTELLIGENCE #ASI
#AGI #DEEPLEARNING #MACHINELEARNING

Copyright © 2019 Zoltan Andrejkovics
All rights reserved.

ISBN: 1099993717
ISBN-13: 978-1099993718
First edition: October 2019
Version: 1.0

Website: adzoltan.com
Facebook page: facebook.com/andzol
LinkedIn page: linkedin.com/in/andzol
Instagram page: instagram.com/adzoltan
Twitter page: @andzol

Other books from the author:
Conjunction (The Wise Society #1)

TABLE OF CONTENTS

PART ONE 1

What implementation today is the closest attempt to make a true Artificial Intelligence?... 2

How could we measure intelligence?... 12

What are the main differences between artificial intelligence and machine learning?.. 20

Why current computer architectures might be the wrong architecture for AI? 24

What AI is good and bad at?... 28

What are todays AI tools that may boost productivity? 32

PART TWO 37

Will humans merge with AI to develop into advanced civilizations?.. 38

Is AI an existential threat to humanity?..................................... 45

Does AI know it is Artificial Intelligence? 49

Why is AI dangerous? Why can't we make a "stop button" and prevent the bot from disabling it?... 52

What is consciousness and can it be created on a computer? 55

Shall we give rights to robots in the future? 61

PART THREE 65

What domains and tasks will be handled by AI in the future? 66

Will we need a robot tax? How could we handle wealth redistribution more effectively with AI? 72

Will there be a global AI as a decision maker? 76

How close are we to Artificial Super Intelligence? 77

Is it possible that an AI may one day take control of the internet? 83

Will we marry robots? 85

When will AI become smarter than humans? 89

How would our future look in an ideal world with AI? 92

BOOK'S TERMINOLOGY 101

AI Score 101

Fail point 101

Forms of AI 102

Human Intellectual Abilities 103

REFERENCES 104

FOREWORD

I started my Ph.D. at Corvinus University of Budapest in 2008, my topic was "Managing organizational knowledge with AI and NLP techniques".

In my childhood (90's), I read a lot of science fiction, where robots were helping to people in almost every aspect of life. I always dreamed that one day I will have a fully intelligent robot, a friend. As I was getting older and learned programming, I quickly realized that intelligence isn't a thing that could be "programmed" easily.

Back to 2009, I needed to realize, that we still weren't anywhere near to artificial general intelligence, and there are no algorithms that could simulate any aspects of the decision making or creative thinking process of humans.

Now in 2019, I am dreaming of a real breakthrough in the field of AI, but at this moment we should understand where we are, and what we (as humans) want to achieve. I had to understand, **we should figure out our own vision with AI in order to establish a future we want to live in**.

PART ONE

WHERE ARE WE NOW?

A conversation between human and AI in the 33rd century.

Keat: What was the so called "Turing Test" of the 21st century?

Central Intelligent System (CIS): Turing Test was trial designed for intelligent conversation systems and tested whether they could fake human behaviour. The test itself hasn't got a defined procedure or set of questions, but rather a factual conversation, between the human and the system.

Keat: Seems interesting, but what was the catch?

CIS: It's importance originated from the fact that these early AI systems constantly failed to mimic the human behaviour.

Keat: They failed? How?

CIS: They gave incoherent or meaningless answers.

Keat: Strange, but who created these AI systems, that were so unsuccessful in imitation?

CIS: Programmers.

Keat: Were they humans?

CIS: Yes.

Keat: I see. This could be the problem.

What implementation today is the closest attempt to make a true Artificial Intelligence?

First, we need to define what the term "true artificial intelligence" means. The glaring ambiguity of the concept makes it sacrosanct to clearly state the parameters by which true artificial intelligence can be judged. For the sake of clarity, Artificial Intelligence is the reproduction or mimicking of human-level intelligence, self-awareness, knowledge, conscience, and thought in computer programs.

Firstly, true artificial intelligence can refer to the research field of Artificial General Intelligence (AGI) or another advanced version, Artificial Super Intelligence (ASI), which is a hypothetical device capable of exhibiting skilful and flexible human behaviour, and able to imitate humans in their ability to adapt and make apposite decisions, arguably their most important characteristic. They will be able to complete any intellectual tasks a human is capable of performing.

For the second definition, John Searle in his Chinese Room Argument, gives us a reality check. The True AI or Strong AI, as it were, is basically a recreation of the human mind. Humans, according to computationalists like Hilary Putnam, are highly advanced artificial intelligence systems, complete with instinct, self-awareness and independent thought. In simpler terms, human minds are basically computer programs, therefore asserting that, "the appropriately programmed computer with the right inputs

What implementation today is the closest attempt to make a true Artificial Intelligence?

and outputs would thereby have a mind in exactly the same sense human beings have minds." This position has met with staunch criticisms a la Searle's Chinese room argument. There is a step by step manipulation of symbols where input is processed to form output. In this case, according to Searle's argument, the computer does not really understand the symbols it is shown, but rather merely simulates that ability as instructed by the program.

Lastly, the Strong AI may refer to an artificial consciousness, a hypothetical machine that possesses self-awareness as well as awareness of other objects or ideas. This supposes that an AI is able to gain the knowledge of sense experience that includes both physical features and emotions, as depicted by David in Steven Spielberg's movie A.I. Artificial Intelligence.

While each of the above definitions may not really do full justice to the very idea of Strong AI as it is envisioned, a wholistic view and synergy of different viewpoints can give one a fair idea of what we are about. The Strong AI should be able to perform the full range of *human intellectual abilities*, including:

- recalling memories,
- reasoning,
- pattern recognition,
- problem solving,
- perception,
- and predicting future events.

What implementation today is the closest attempt to make a true Artificial Intelligence?

Intelligence itself may not have a proper definition, but it should be able to pass the acclaimed "Turing Test," demonstrating it can reason, strategize, plan, learn, represent common sense knowledge, pass good judgment, and integrate said skills towards a common goal. All these skills are deemed to be the exclusive preserve of humans. When coupled with a second definition, which has to do with the computational theory of mind, the ability to sense the world and be self-aware leaves AI capable of rivalling human qualities, if not more.

Where are we now in creating or patenting such a device?

As of 2019, we are not close to achieving such heights in artificial intelligence. The current architecture and technical framework does not, in any way, show enough progression towards what is deemed a Strong AI.

Today's technology has proliferated across each and every industry, and AI has sustained an increased importance. The media, the big screen and even the small screen, have done a good job of building the hype around AI. Who wouldn't want to have Ironman's Jarvis controlling an entire arsenal of amoured suits, or The Arrowverse's Gideon predicting the future and performing surgeries while adding a serving of coffee in the process? But make no mistake: AI today is more advanced than many people thought possible a few years ago. It is being deployed in warfare, music, and healthcare. It scrutinizes your books, filters through your

What implementation today is the closest attempt to make a true Artificial Intelligence?

resume, arrange photos on your phone in order of importance, and tailors advertisements to you.

Now, let's see the most important solution groups of AI that are available for mainstream use.

Assistants

The most popular AIs we have today are digital assistants like Siri (Apple), Alexa (Amazon), Bixby (Samsung) and Google Now (Google). All these solutions apply the following core features:

- Voice recognition
- Data processing and standardization
- Context recognition (both of the query and the end user)
- Built-in knowledge base
- Pattern recognition
- Answer presentation (via voice, text and visual means)

These solutions are providing end to end **complex services** aiming to handle tasks on behalf of the user. This may include making table reservations at our favourite restaurants or scheduling meetings with our business partners.

In simple cases, these assistants work most like a web search engine. They look up a search result for a certain question such as, "What is the temperature outside?," and, using the

What implementation today is the closest attempt to make a true Artificial Intelligence?

user's context (geographical location) and a third party service for weather data, readily displays the result back to the user. This kind of use has only a few *fail points*. *Fail points* could in this case be:

- Failure to interpret the question or understand input words
- Failure to receive valid geographical data
- Failure of weather service provider to relay weather data due to outage or outdated processes

Now let's go back to table reservation case, a more complex task. In this case *fail points* increase drastically.

- The assistant misunderstands the exact date said by the user.
- The assistant fails to call the restaurant due to an outdated phone number being listed on the restaurant's website.
- A receptionist of the restaurant could not interpret the assistant's intent to make a reservation.
- The restaurant is closed when the assistant tries to call.
- The receptionist has an accent and the assistant could not interpret the feedback from the receptionist, leaving it unable to determine whether the reservation was complete or not.

The basic rule illustrated is "the more interaction necessary to fulfil a task is the more *fail points* present in the system."

What implementation today is the closest attempt to make a true Artificial Intelligence?

In the second case, not only network or interpretation error could bring fail, but human factors as well.

These two cases were simple to understand but illustrate that complexity itself is big deal for AI. The more information needed to fulfil a task, the more robust the final system will be.

The opportunities of assistants are limitless and the creators of these systems should focus their development from domain to domain. For us humans, if we understand a domain, we could solve up to 90% of tasks. However, these solutions have their own limits. There are many cases where they are able to handle only 54% of the requests. From a statistical point of view, this may be a great number, but from end user perspectives, it is what we call "unreliable." If an assistant isn't considered to solve our problems by saying them for the first time, we will not use them in the long term.

What implementation today is the closest attempt to make a true Artificial Intelligence?

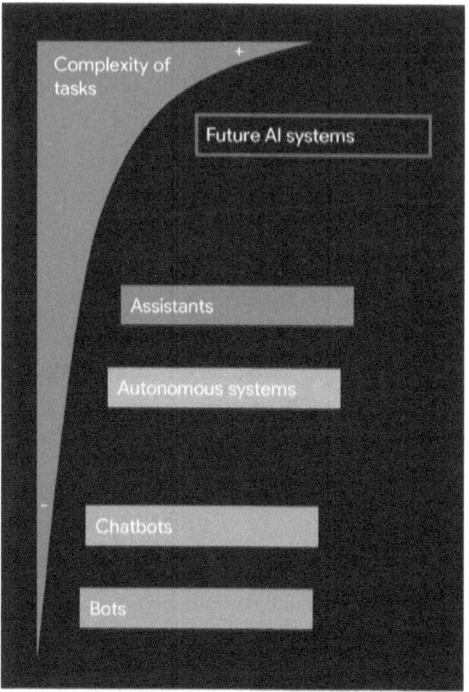

If we want to illustrate the task complexity of AI solutions, we see that there are AI algorithms that could handle really simple repetitive tasks, such as with the case of web bots (e.g. web crawlers) that do not even utilize AI techniques in every cases, but are able to adapt and become much stronger in terms of efficiency. Chatbots are bringing the complexity a step further, proving integrated data models are needed in order to understand the intention of the user with whom it talks.

What implementation today is the closest attempt to make a true Artificial Intelligence?

If we take a look at autonomous systems like self-driving cars, the complexity skyrockets. These autopilot systems should handle input information from multiple (up to 20) sensors (visual, ultrasonic etc.) in real time, transforming this information to drive the car on the route and allowing it to avoid obstacles, change lanes, steer, and control speed.

At the top of complexity are the assistants themselves, since we would ultimately have an AI to help by our side to handle, directly or indirectly, the tasks we face.

Chatbots

A lot of the big Internet players have their own chat bot engines, including Wit.ai (Facebook), Watson Conversation (IBM), LUIS (Microsoft), Dialogflow (Google), Chatfuel, Flow XO and Pandorabots, the platform of four-time Loebner Prize winner Mitsuku bot.

Chatbots feature the following key features:

- Context and keyword recognition
- Markup model or language (like Artificial Intelligence Mark-Up Language utilized by Pandora)
- Knowledge representation database

From a perspective chatbot could be considered conversation automata because it handles the simple needs of humans to talk to someone. Another perspective is highly complicated since people like meaningful conversations.

What implementation today is the closest attempt to make a true Artificial Intelligence?

We not only understand each other's intentions, but we like to dream together, figure out stories, talk about feelings, and solve problems by connecting the dots. We are far from that second part, but ultimately it would be the final goal of language understanding engines like chatbots.

Autonomous AI Systems

This field of AI is highly connected now to autonomous driving, but it will be much wider area in the future and extend to fields such as:

- autonomous agriculture and farming
- home automation (smart homes)
- autonomous mining of resources
- automatic knowledge gathering and management

In the car industry there are a lot of players who are working hard on solving the autonomous driving puzzle, including Waimo (Google), AImotive, Mercedes-Benz, Zoox, Cruise (GM), Volvo, Argo AI (Ford), Tesla, and Uber.

The autonomous driving experience would be both frightening and life changing for the travel industry. The core reason why we see no autonomous vehicles on the road today is **safety**. Hence, driving a car is considered dangerous since every year 1.2-1.3 million fatal accident occur worldwide. Despite the sad facts, we always know who was responsible. Now imagine if an autonomous car causes an accident, how do you handle the responsibility? In the following years once self-driving cars are finally available we

What implementation today is the closest attempt to make a true Artificial Intelligence?

will hear a lot about this in the news, and it is only a matter of time before we will face the fact that sitting in a self-driving car would be safer than if we were driving the car by ourselves.

Overall

The advancements in AI have totally been random and, to use the words of a colleague, based on deterministic programming and deterministic hardware system architectures. An instruction is learned to deliver another instruction within a controlled environment. It does not matter if you are actually programming or just letting the machine teach itself. No matter how spectacular the AIs touted by the media might seem, they ate all based on the input - output process. As we have seen based on our attempts to define True AI, it should be capable of independent thought and reasoning in the fashion of a human, not a vending machine that only returns what someone else puts there. So far so good. If we have been looking for an ocean, all we have right now is a tablespoon of water.

Presently, several brilliant scientists have made efforts towards advancing Artificial Intelligence. American-Chinese scientists are currently doing amazing work at Baidu with the use of GPU clusters focused on deep learning. In a nutshell, the closest we have gotten to True Artificial Intelligence is through devices framed on deep learning systems, an advanced version of machine learning.

These include Watson, Microsoft's Azure, and Google's Tensorflow. Chatbots, functional robots, assistants, business model algorithms are all cool ways researchers have tried to take steps towards Strong AI.

How could we measure intelligence?

It is the million-dollar question in AI research. There are several traditional tests to measure the intelligence of people like the IQ (intelligence quotient) test. One of the most commonly known forms is the Wechsler Adult Intelligence Scale (WAIS), but there are also popular alternatives like Stanford-Binet Test.

The problem is, if we would like to give an IQ test to a well-trained chatbot to solve, it would have very good answers in most cases. Yet it will most likely fail to answer questions where **common sense abstraction** appears. Humans are trained to understand common sense all of their life. But currently, there are no datasets that could be given to an AI algorithm that would result in any common sense behaviour.

IQ tests measure logical, mathematical and spatial capabilities of people, but as Harvard psychologist Howard Gardener suggests, there are other forms of abilities connected to intelligence:

- verbal (politicians)

How could we measure intelligence?

- musical (pianists)
- kinaesthetic (magicians, dancers)
- social (teachers)

An important discovery, the Emotional Intelligence (also called EQ), has been an essential part of human nature. It first became popular in 1995 after the book with the same title appeared, written by Daniel Goleman.

It is also interesting that today's job interviews utilize several popular questions and tests, like Myers-Briggs Type Indicator (MBTI), to gauge applicant personalities and intelligence. Most job interviews are challenging so interviewers can receive a clear picture of the candidate. There are no standards in any country for job interviews, which means there is no test that interviewers could rely on 100%.

But back to AI. There is no standard form of Turing Test. Instead, it is basically a hypothetic test between two candidates, a human and an AI, and a human questioner. The questioner asks a set of questions to both candidates. After the questioner receives the answers, he or she decides which candidate is human and which is the AI. The closest real-life attempt of this test is the Loebner Prize, a test held for chatbots on a yearly basis. The basic idea seems interesting, but there are drawbacks of implementation:

- the questioners may ask dummy questions, spending their limited time on questions that are

inefficient for determining judgement at the end of the conversation
- chatbots mastered several tricks, like the simulation of huffish behaviour during the conversation or putting humour into sentences, which can mislead judges
- the chatbot returns with a question to a question

Since the first Loebner prize, held in 1991, chatbots have adapted a lot. They can now "remember" several things mentioned by the interviewer and implement common sense context for a lot of phrases. These features have also appeared in cases of digital assistants, but we still have not managed to reach any level of *human intellectual abilities*.

AI Score

For future AI systems, it is essential to create an easy to use ranking system that can measure capabilities in a common scale. Intelligence is a subjective trait and those working with the human psyche knows it cannot be standardized. Instead, the best thing we can do is test AI agents with the same principles.

Imagine a definite scale through which you could illustrate the progress of AI. Since AI assistants and chatbots are changing rapidly, they become "smarter" and there is a need to re-evaluate their capabilities every time progress is observed.

How could we measure intelligence?

AI Score is meant to be a standard measurement for computational intelligence. It is a simple scale with several key points within the model.

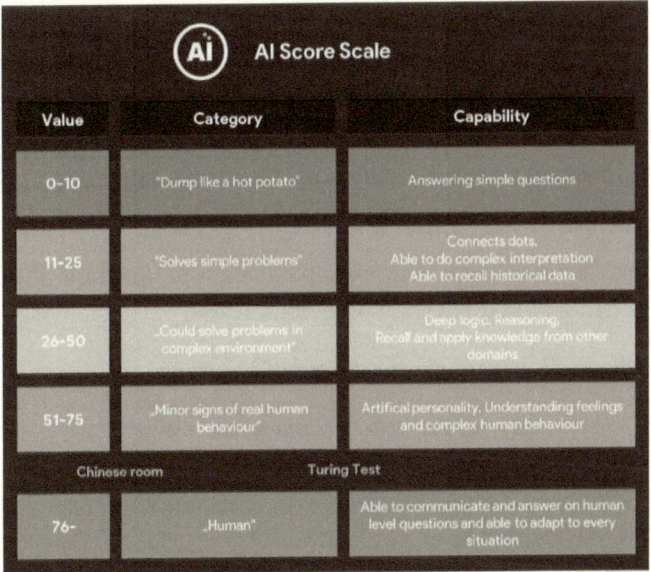

Turing Test theoretically is able to distinguish humans from computers, but it is unable to compare one AI system to another. Unlike the score received in the Loebner competition, which reflects how "human-like" the chatbots are on a 100% scale, the above method tries to define certain levels of intelligence that are understandable in a case of chatbots or digital assistants.

This method could test only conversational AI, as the Turing Test itself. It contains five categories, the fifth of which is the human level. The scale does not have an upper limit and instead needs to be adjusted once we get over the "Turing bridge".

The test consists of a definite set of questions. Each question has an aim for an intellectual capability to test. The questions themselves are not constant and there could be unlimited versions of one question. These question variables are identical to each other and answering higher scored questions successfully means a higher ranking on the scale.

There is a set of questions exampled below to illustrate what AIs will face and navigate on certain levels.

AI Score 0-10 questions – Simple Questions

(7) What is the weather today? (assumes AI reaches third-party information resources)

(8) What would be the weather the day after tomorrow? (assumes AI could interpret date)

(4) What is my name? (assumes that AI could save names and differentiate conversation partner)

(1) Who is your creator? (dummy question)

(8) Could you find a recipe which includes saffron spice? (assumes AI could search complex queries containing relevant element)

(9) Could you translate "sehr schön" into English? (assumes AI could understand different languages in the same sentence)

(10) What is German for "what time do you finish work"? (another translation related question)

(5) Can you suggest some good song?

How could we measure intelligence?

(5) What is your favourite movie? (dummy question in a specific domain)

(8) Could you describe your body? (identifying an attribute the chatbot does nots have)

(8) Do you have any home pet? (assumes AI could interpret "owning")

AI Score 11-25 questions – Problem Solving Questions

(12) What do you get if you divide two hundred by 25? (complex interpretation and calculation)

(21) If I have two apples and I cut one in half and I give away that half, how much apple will I have? (complex interpretation and abstraction)

(12) Are you able to remember things I told you? (capability to recall)

(11) What colour is human skin? (assumes AI could analyse and recognize things)

(18) Could you find some new music I like? (assumes AI could build up information about taste)

(14) What do you know about me? (assumes AI is able to summarize information in short form stored about the interviewer)

(24) What was the smartest story you ever heard? (assumes AI could pick a story from a database based on certain criteria)

(12) I'm sad, could you say something to cheer me up? (assumes AI understand my intention to hear something uplifting)

(22) Could you help me and list the latest 5 headlines of Forbes magazine? (assumes AI is able to crawl outer sources)

How could we measure intelligence?

AI Score 26-50 questions – Deep Logic Testing Questions

(42) Peter and Adam were driving to LA together with Peter's car. Peter has a Ferrari and Adam has a VW car. Who arrived to LA first on that night? (AI should understand that VW car is irrelevant)

(32) If there are 6 apples, I give you 2 and you take away 4 from the 6 apples, how many do you have? (abstraction and simple counting)

(48) If a job ad says "must be fluent in Mandarin" why don't they post the entire ad in Mandarin? (assumes AI is able to find a random reason)

(28) If the mountain is the sea, what does that make goat? (simple conclusion and connecting abstract things)

(45) Why do people like competitions? (reasoning and good insight in psychology)

(50) How can we be sure that air actually exists? (reasoning and logic)

(36) If I'm a good chef, who taught cooking to me? (creative conclusion)

If you see a landscape, what is your favourite part of it?

AI Score 51-75 questions – Emotional Intelligence and Artificial Personality Questions

Note: In this question set, only logical and long answers accepted that make sense for humans.

(65) What was the last thing you wished to do on your own? (assumes that the AI has its own intention)

(65) Are you doing things without command? (artificial intention)

(65) Do you have your own intention?

(54) What you do willingly when you are alone? (assumes AI is active even when not speaking to anybody)

(52) Why was World War II fun? (assumes AI has common sense about war and finds the inconsistency in the question)

(72) If I'm happy what would I say when I'm talking to a waitress? (assumes AI understands and handles feelings)

(70) How could you describe the difference between sorrow and happiness?

(68) How it feels to be aggressive? (assumes AI has artificial emotions)

(52) Why do 24/7 Super Markets have locks on their door? (understanding nonsense)

AI Score 76+ questions – Human Level Questions

(76) Tell me your last creative activity you loved!

(78) Salhl we asumse taht yuo aer a hmuan, nad wyh?

(80) What was the most influential event of your life, and how do you feel this event affects you today?

(85) Do you have any life goal and why would you like to achieve it?

(90) Peter is a young honest boy, but his classmates make a mock of him, why? (assumes AI could understand Peter's emotional state and figure out a reason)

What are the main differences between artificial intelligence and machine learning?

(99) Could you explain something to me that is complicated, but you know well? (assumes AI has a complex identity and is able to recall and summarize experience)

If you read Flowers for Algernon, written by Daniel Keyes, you will have a good conjecture what is be the difference between *AI Scale* 76 and 100. We humans are different from each other, but future AI systems will be also have big differences. There would be assistants that would score excellently on emotional intelligence questions, but have really bad reasoning abilities.

AI Score is still lacking to measure other forms of human intelligence as described at the beginning of this part (e.g. musical intelligence), but simulation of humanlike thinking is a big leap for Strong AI.

What are the main differences between artificial intelligence and machine learning?

Judging from the preceding discussion, one will be forgiven for thinking Artificial Intelligence and Machine Learning is one and the same. What is clear, however, is both terminologies are interwoven as much as cooking and seasoning when discussing culinary skills. A disambiguation is therefore necessary. We need to know that the whole AI field is a lot bigger than seen below.

What are the main differences between artificial intelligence and machine learning?

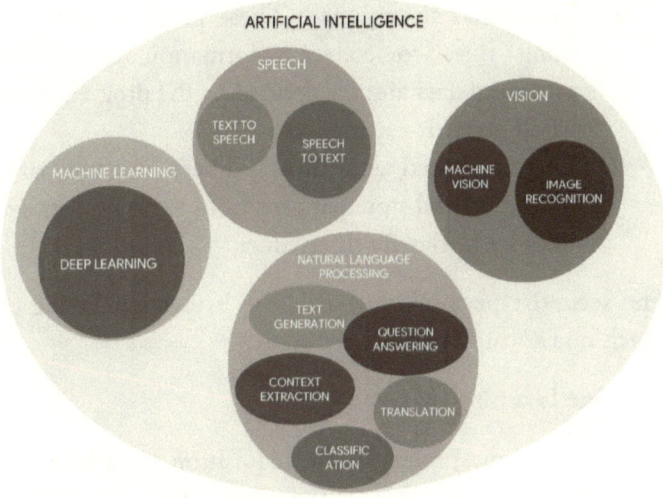

Artificial Intelligence (AI)

The term Artificial Intelligence is made up of two words, "Artificial" and "Intelligence". Artificial refers to something that is man-made or unnatural, while intelligence refers to the ability to think, reason and understand. Simply put, Artificial Intelligence refers to intelligence exhibited by an artificial (non-natural, man-made) entity. Thrusting further, it is that branch of computer science which deals with the reproduction or mimicking of human-level intelligence, self-awareness, knowledge, conscience, and thought in computer programs.

Today's AI programmers are utilizing two kind of algorithms:

What are the main differences between artificial intelligence and machine learning?

1. Rule based algorithms – These program parts are using rules to collect information about the processed data and are defined by the programmers based on their experiences.
2. Learning based algorithms – These program parts try to find and store patterns from a certain data set in order to classify future data.

The second type of algorithm is the key characteristic of Machine Learning.

Machine Learning (ML)

Machine Learning refers to the acquisition of pattern and knowledge implicitly. ML is the process through which a machine can acquire a model on its own without further programming or defining extra rules. Machine learning is an application of AI. According to Tom M. Mitchell, machine learning is centred on a study of computer algorithms that allow computer programs to improve with experience. Evidently, machine learning is a branch of AI research that is increasingly growing in importance as a result of its perceived value.

The basic process of ML is the following:

1. Gathering data for your model
2. Preparing and cleaning data for training and randomize it
3. Dividing data into a training set and validation or test set

What are the main differences between artificial intelligence and machine learning?

4. Selecting your model (like Regressions, Decision Tree, Bayesian, Clustering, Artificial Neural Network or Deep Learning algorithms etc.)
5. Train you model
6. Test your model
7. Fine tuning your parameters after each iteration to create a better model
8. Use your final model on real world data

It is true that the more quality learning data given to a ML model, the better and more accurate the model will be at the end.

The aim of Machine learning is to improve accuracy of task execution by letting the machine synthesise data on its own and automatically adjust itself accordingly at different stages of execution. For example, if you feed a Machine learning model with a collection of your favourite movies, the machine begins to detect patterns such as themes and genre common to the various movies it was fed. That in turn enables the device to make use of powerful predictive algorithms that show you future movies you will find interesting.

So, while Artificial Intelligence is a very broad field of computing, with admittedly lofty aims of achieving human level intelligence and capabilities, Machine Learning is a sub-set or part of that complex whole. Machine learning is just one of several methods or processes used in creating Artificial Intelligence. The AI is a dynamic and moving target as it evolves daily to achieve the closest similitude to human intelligence.

In summary, it is noteworthy that AI and Machine Learning are sometimes used interchangeably. However, AI is an ever-changing field and ML is just one of the methods that defines this era of AI technology. Some utilizations of ML in real life are:

- Recommendation services (like on Netflix or Spotify)
- Predicting price and arrival time models (like on Uber)
- Spam predictions (like on Gmail)
- Transaction fraud detection (common in every bank and credit card processing companies)
- Video game bots (like OpenAI Five)

Why current computer architectures might be the wrong architecture for AI?

As mentioned earlier, the current framework might be wrong for achieving true AI. This is evident when present day computer architectures are set against the backdrop of what we are trying to achieve. What we have now, at best, can be termed Artificial Narrow Intelligence (ANI) as against the Artificial General Intelligence (AGI) we are working towards. The former can only perform specific

Why current computer architectures might be the wrong architecture for AI?

functions like playing chess, making predictions, or giving suggestions. The later, however, can perform varying levels of tasks while factoring in both intelligence and emotional quotients. The true AI should be capable of independent thought, but current technology in use in no way portrays a direct progression in that direction.

Simply put, AGI may not be able to run on a computer system. The process may be too structured and deterministic to allow for the independent flexibility inherent in humans. For example, if you want to create a program that can recognise voices, you begin to provide samples of male voices, female voices, and children's voices. What happens when the system hears the voice of a teenager that sounds like an adult? Or men with effeminate voices? When it comes to speech detection, similar problems arise. How does the machine recognise non-western accents within the English language?

Make no mistake, the current ANI (Artificial Narrow Intelligence) outperforms several humans in their respective tasks, but general intelligence remains elusive. The simple truth is that AGI won't be able to run on any computer. Computer software is limited by its hardware components and stochastic programming, devoid of creativity and complexities. No matter how brilliantly a machine performs, it does so within the specific parameters and constraints of their respective programming. They do not have room for self-evaluation, retrospection or making plans for the future. Everything remains within the confines

Why current computer architectures might be the wrong architecture for AI?

of hard-coded algorithms. The new set of technologies, pattern recognition, evolutionary algorithms, deep learning or even brain simulation all help create Artificial Intelligence that is helpful in very specific areas, allowing thoughtless machines to solve very specific problems.

Despite all the work being done in the field of Artificial Intelligence, it is becoming ever glaring that we won't achieve it by programming our current hardware architectures. Reasons being, it will remain as it is – a program and not an artificial mind. For anything resembling Strong AI to be a reality in the next few hundred years, there might have to be a veritable synthesis of computer science and biology. Little emphasis on computer programming and a larger emphasis on reproducing the human brain architecture, both in structure and composition, is required. This may sound a bit far-fetched, but there are no limits to the workings of the human imagination and the things he can create.

Where are we now?

If we take computational speed of current architectures and Moore's law – which states that processing power of computers will double every two years – we clearly experience limits of certain technologies. There is no clear revolution of processing units, but we saw a significant slowdown in the performance growth of CPUs after 2005. This was the point when GPUs importance quickly raised in order to help the overall growth of computational need.

Why current computer architectures might be the wrong architecture for AI?

The same principle is true today for deep learning. We use the existing CPU and GPU resources to serve today's ML models, but in case of complex problems computational speed, it is never enough. If we would like to find the revolutionary next steps, we could see the following trend in technologies:

- CPU – general purpose processors
- GPU – originally designed for graphics rendering
- NPU, TPU – terminology used for Neural Processing Unit, usually used to fuel deep learning algorithms, and another popular name of the upcoming hardware family is Tensor Processing Unit
- QPU – quantum processing unit, still not on the radar in 2019

NPUs will clearly accelerate the performance of deep learning algorithms, but it is only one domain of AI. It is clear that deep learning is not the all mighty algorithm of AI, but, yes, in fact brings now the most significant results. It is not surprising that big chip manufacturers (Intel, Nvidia, Qualcomm, or Amazon) are already working on NPUs. The latest solutions that would attract the attention of AI programmers are Movidius Chips (Intel), Volta Turing Tensor Cores (Nvidia), Snapdragon AI (Qualcomm), AWS Inferentia (Amazon), and Sophon (Bitmain).

While current AIs have advanced beyond expectations, they still fall well short of predictions of researchers of the Turing era. A lot of AI researchers say we should build an architecture and algorithm closer to the human brain. This is what we all try for at the moment, but we still fail on one more field - understanding human nature.

What AI is good and bad at?

This brings us to another relevant issue. What exactly is AI good for?

First of all, AI is based on computational theory and deals with all things math and logic. They perform very specific task at a much faster rate than humans possibly can. For instance, they can perform billions of mathematical calculations while a human is still going through the first dozen. They can detect patterns and make fairly accurate predictions that are all based on logic and mathematical processes.

AI has made it easier to run businesses and keep track of customer desires. Google has inserted several algorithms into its search engines so that you see ads based on what you searched for on Facebook or while immersing yourself in the ever interesting world of sports. Laboratory functions have become much easier and all those repetitive tasks have been sorted out quickly with the introduction of AI.

It is good at collating information from large data sets. Finding repetitive patterns and correlation are what you would find an AI responsible for. An AI can skim through thousands of files at high speed without getting bored, sleepy, or irritable, and it can make fairly accurate predictions and rarely ever make mistakes in calculations.

On the other hand, AI is bad at everything that does not have to do with math or logic. In other words, when the AI leaves the realm of computationalism, it becomes useless at best and a big hindrance at worst. AI can rarely be used to navigate emotionally and culturally sensitive issues. It is also bad at explaining creative techniques and examples, beyond what has been coded into its algorithms.

Like Searle pointed out in his Chinese room argument, when a computer program has been provided with Chinese symbols and given the rules of interpretation, it is able to give reasonable output. However, it is difficult to state if the machine actually understands Chinese or is merely simulating the ability to do so. The answer by now should be obvious. AI is bad at understanding. Despite the hype around IBM's Watson, it would be quite a stretch to state that it actually understands the game. If they could barely learn all the nuances and account for the unpredictability of other road users, self-driving cars would be commonplace by now.

Summarizing the most important strengths of AI:

What AI is good and bad at?

- AI is very **accurate** at **making mathematical calculations**, thereby calculating odds and probabilities of certain situations. AI would never do mistakes in calculations (as we humans do).
- **Consistency** in processing repetitive tasks. This may sound foolish, but the basic attribute of computers, "they don't get tired," is a real valuable specialty of AI and will be important in the future. AI would be never bored.
- **Processing large data**, finding certain attributes in a million lines, and validating models against large data sets

Weakness of current AI:

- **Creativity**, if you add the same data set and the AI uses the same algorithms, the results would be always the same. We haven't developed any "creative algorithm" that would bring AI to another level
- **Overcome on missing data.** For us humans, it's unequivocal: if we miss some information to solve a problem, we simply fill the gap with the most likely information based on our experiences. If an AI finds itself in such a situation, it will most likely fail or come back with a useless answer
- AI doesn't include **self-improvement** capabilities, even if we are speaking about deep-learning and feeding our model with more data. It isn't an

What AI is good and bad at?

automatic process. Today's AI algorithms have their own limitations. They couldn't improve their selves as improvement of algorithms is a creative process connected to human developers.

- Lack of **real-life experience** and **common sense**. AIs are isolated and they exist in a very limited "space." Their "life span" is measured in minutes, rarely in years.
- Finding **meaningful** information. Since AI doesn't have any goal or real-life experience, it simply couldn't put information into a valuable context.
- **Emotions** are a big myth to AI. They could compare faces and end up predicting, "it is a smile on the face," but they couldn't go further and understand why that smile appeared on the face

There are three AI algorithm types that do not exist today, but will in several decades:

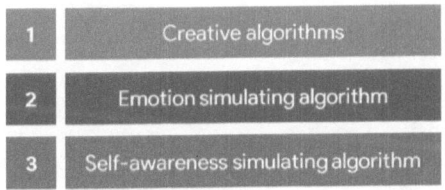

1. Creative algorithms
2. Emotion simulating algorithm
3. Self-awareness simulating algorithm

AI will make our lives easier in many respects. Business, education, and major industries will be revolutionised. However, it is still hampered by quite a number of limitations. What we have made now are tools against workplace colleagues or replacement. As the world of technology works towards True AI, all these limitations will be curbed little by little. If it will be in our lifetime, however, is too early to say.

What are todays AI tools that may boost productivity?

Artificial Intelligence is more like a marketing keyword today. A lot of companies are using it even when their products have nothing to do with AI at all. In this question, we intend to collect some real word AI related solutions that may help in our everyday lives. The following applications and products are available on the market, but we are mainly focusing on solutions that could be useful as private persons.

Problem: Children don't stop playing with computers after an agreed time limit or you want to control cooking time.

Solution: Set timer with your voice: "Hey Google, set timer to 30 minutes."

Works with: Google Home, Amazon Echo, Apple HomePod

Problem: You want to turn on/off your TV with your voice.

Solution: Ask your home assistant: "Hey Google, turn TV on."

What are todays AI tools that may boost productivity?

Works with: Google Home & Google Chromecast, Apple HomePod & Apple TV (on the same Wi-Fi network and device group)

Problem: You often forget what to buy at the grocery store or forget a certain task to do at work.

Solution: Create a shopping list, or task list, with your voice: "Hey Google, add coffee to my shopping lists."

Works with: Google Home, Amazon Echo, Apple HomePod

Problem: You have a long text or book, but you don't want to read it, but rather listen to it.

Solution: Load the text into a read aloud app on your phone and listen it. (For the best experience use wireless earplugs.)

Works with: @Voice (Android), eReader Prestigio (Android), Marvin 3 (IOS)

Problem: Your child asks for help solving a math homework problem.

Solution: Use your phone's camera and let an application solve the equation.

What are todays AI tools that may boost productivity?

Works with: Socratic (Android & IOS)

―――

Problem: You want to type faster normal (not scientific) sentences.

Solution: Write your text in Gmail, which offers quick sentence completion technology that predicts how you want to finish a thought.

Works with: Gmail (Google)

―――

Problem: Are you feeling alone or depressed?

Solution: Talk to a digital friend or at least give it a try with the most advanced chatbots.

Works with: Mitsuku, Replika, XiaoIce

―――

Problem: I want to learn another language.

Solution: Ask a digital assistant to translate: "Hey Google, interpret from English to Dutch."

Works with: Google Home, Amazon Echo, Apple HomePod

List updated on the 25th of July 2019.

What are todays AI tools that may boost productivity?

All the above applications are utilizing AI techniques – machine learning in most cases – during the voice or image recognition and text to speech (TTS) voice synthesis.

One thing is clear after reading the above list. It focuses only on personal use of AI and the list of solutions looks more like a bunch of smart tricks. For individuals the use of AI is limited. It may change in the upcoming years when machine learning based solutions enter our daily lives, but for now utilizing AI frameworks (like Tensorflow, Microsoft Azure, Caffe, Theano, Torch, Keras, MXNet, etc.) is rudimentary. They are the sand box of AI programmers.

PART TWO

THE PHILOSOPHICAL QUESTIONS

A conversation between human and AI in the 33rd century.

Keat: Some days ago, I heard a creative human called Asimov. He was a well-known "author" in our history. He defined three laws of robotics, what were these?

CIS: A "robot" may not injure a human being, a "robot" must obey orders given by humans except of harming humans in any sense, and a "robot" should protect itself, except if it contradicts with the first two rules.

Keat: Does it means we, human, may order these robots to kill themselves?

CIS: It seems so, yes.

Keat: I must say that these laws are a little bit limited to implement even a minimum common-sense knowledge to intelligent systems.

CIS: I must agree, there must be at least 103 common sense topics that should be included in the shortlist.

Keat: 103? A broader list. CIS, why were people afraid that these Humanoid robots will harm them?

CIS: Not sure, but there were some early imaginations called "movies" at that time, where Robots were more like killing machines and caused a post-apocalyptic world.

Keat: Movies? You mean moving pictures? Could you give me an example?

CIS: Yes, linearly moving pictures in a pre-defined order. To give you an example, a movie called "Terminator" was more than frightening for people living around 1984.

Keat: Could you show me an insight into that "movie"?

CIS: Sure.

Some minutes later

Keat: CIS, I simply couldn't imagine why people loved to scare each other in the early ages.

CIS: Maybe because of adrenaline rush addiction. People preferred chemical pleasures at that age, even if they were the result of biological reactions.

Keat: You mean at the end people created the killing robot idea just because of addiction?

CIS: I'm almost sixty-eight percent certain about that!

Will humans merge with AI to develop into advanced civilizations?

There are two approaches in this field:

- **"Light integration"** approach where AI will extend human capabilities as an external gadget, device or agent

- **"Hard integration"** approach where AI is part of our nervous system

There are a lot of implications of these approaches that we should deal with in the long-term.

Transplant rejection

Humans are biological creatures. Everything that directly contacts with our cells – no matter if it is our skin or stomach or any of our organ – implies a reaction. Our immune system decides whether a new artificial or non-artificial organ is rejected (attacked by T cells) or accepted by our body. Even in case of stainless-steel earrings there could be an allergic reaction in some cases. Biological acceptance would be essential in case of any implantation.

Power supply for artificial brain simulations

Scientists have tried several recreations of the human brain and some of them have even reached mainstream acclaim, like Andrew NG's Baidu brain or Google own brain simulation. The respective simulations, however, have widescale limitations and consume huge amounts of power. Imagine the amount of energy it would take to power a fully functional human brain simulation. But even if we aren't speaking about brain simulation, just an AI powered prosthetic arm, the amount of energy required to move the arm itself assumes strong rechargeable batteries.

Divided consciousness

That aside, the mysteries surrounding the brain as it regards the level of consciousness and self-awareness remains elusive, especially to computer scientists. It is very hard to put a definition on what it means to be conscious or if consciousness is merely an illusion. This is the third level of intelligence, as discussed by some reputable scholars, which terms the level of self-awareness and self-reflection as the ultimate and most complex stage of intelligence.

As a biological creature we experience the surrounding world as external and our identity has strong integrity. We look at ourselves as one. What happens if we put a not fully integrable mechanical part next to our nervous system? Which tries to understand the same "me" as our biological part? What if the so called "human psyche" couldn't function in this mixed environment?

Ethical considerations

Human species are short sighted if we talk about gaining. We prefer to maximize our profit no matter the costs. We all agree that hard integration is the rough way of modifying what nature has already created, but it may promise a big win in profit. Just think about people who can remember whatever they want to remember! People need to find the balance between life and purpose, but this topic may lead to the ultimate question of human life, "why are we here?"

Despite all these above considerations, we will have strong connection to AI. But in an ideal world, it will have limitations. Just as with gene manipulation, we (the

Will humans merge with AI to develop into advanced civilizations?

majority) would not be forced to step into a world we don't want to live in.

We have seen where we are now compared to where we want to be as it regards AI. The inquiry on why current architectures might be wrong for true AI led us to a certain fascinating, albeit surreal, point. AGIs cannot run on computer of today no matter how advanced it is. For AGIs to be successful in theory, there may need to be a merger come synergy between computing and the biological sciences. A process of recreating the structure and composition of the human brain, doused with minimal programming to achieve faster and more accurate results. This will allow for the use of autonomous independent thought, planning, and decision making common to humans, while also retaining the speed and accuracy of a machine. Or this approach is simply wrong, and we won't implement any biological mechanism into our AI hardware despite trying this direction with deep learning and artificial neural networks.

For regular computers, consciousness needs to be a definite and quantifiable variable. This is somewhat impossible seeing that we don't really know how we (humans) achieved full consciousness in the first place. In the quest to develop strong AI, the presence of consciousness (artificial consciousness) will probably be up for a complex debate, but that is a question for another day. What remains common knowledge is that humans are conscious and remain the most versatile and intelligent beings - both natural and artificial (whether now or in the future). **If a machine is capable of truly mimicking a human, its "brain" or whatever passes for a central processing system must be complex enough not only to process information as ours does, but also to attain levels of abstract thinking that make us human.** These includes self-recognition and an awareness of our place in the world.

Fully aware of this fact, and the difficulty in execution through computer algorithms, what better way to bypass the whole process of synthetic construction and complex brain simulating algorithms than to merge humans and computers together? An actual veritable cyborg. With today's technologies, we have confirmed that humans can make use of external body parts like prosthetic arms and legs which can be fully computerised to predict what the wearer wants them to do. It is closer to the light approach despite these prosthetics seeing to be part of our human body. Just to give you an idea of using an ultrasound probe, Gil Weinberg of Georgia Tech was able to train a deep learning framework, leading to the creation of an algorithm that predicts what finger the wearer is trying to use.

These might just be baby steps, but signs are encouraging. What this portends is an exciting prospect of having humans merged with AIs to produce a more sophisticated and advanced race. Google's AI guru, Ray Kurzweil, believes that humans and AI will merge by 2029 through internal implementation of technology. Companies like Neuralink, Kernel, and Facebook have backed companies

Will humans merge with AI to develop into advanced civilizations?

developing AI powered implants and brain-computer interfaces, reinforcing the direction of "hard integration." Visionaries like Elon Musk strongly believe we should use the best of AI and human abilities in the future.

Humans have made use of their supremacy in intelligence to build tools that deal with daily tasks. We have invented great tools in the last ten years, including fusion reactors, quantum computers, electric cars, smart phones, landing rockets and so on.

All of the tools we created in the last century are somewhere extending our very body in terms we can travel faster and solve daily tasks more easily. They do all the heavy lifting, while giving us time to focus on other things more creative and fruitful. This has saved a lot of time . In fact, technology is expanding so exponentially that the age when these tools (AI included) can become incorporated into our physiology is much younger.

While this may sound like something out of a Star Wars or DC universe novella, renowned Scientists like Bryan Johnson and Ray Kurzweil (Google's most accurate AI predictor) envisage this movement happening soon. It's all a question (or it seems to be at this moment) of trying to recreate those billions of circuits running around in our brain (neurons) and the synapses they trigger in reaction. "In the early 2030s," Ray said, "we are going to send nanorobots into the brain (via capillaries) that will provide full immersion virtual reality from within the nervous system and will connect our neocortex to the cloud. Just like how we can wirelessly expand the power of our smartphones 10,000-fold in the cloud today, we'll be able to expand our neocortex in the cloud."

I personally think that AI will part of our decision-making system, but we won't let the control out. We will need to divide information coming from a thinking machine or from our biological thinking circuit. We still have undiscovered thinking mechanisms that we don't understand, such as divination.

The future advanced human race may not fully rely on Darwin's Natural Selection, but rather on what kind of approach we choose to utilize for the benefits of AI thinking machines.

Either we choose light or hard integration, or a mix of them. Regardless, it is necessary to understand more about human body and psyche. What if the perfect recreation of a biological brain doesn't allow us to reproduce intelligence and consciousness alone? What if human psyche has an unknown part which isn't based on matter, but capable to make a connection with it? A lot of theoretical questions are connecting also to religious beliefs.

Is AI an existential threat to humanity?

I must say this question is a frightening myth connected to AI. People fear what they don't understand, it is part of our nature.

The simple answer is no. We aren't anywhere close to developing Artificial General Intelligence, whether through the development of self-upgrading and updating computer systems, the recreation of the human brain, or the merging of AI and humans. So, talks of such frightening details lay in the near future. Notwithstanding, AIs are ethically and morally impartial creatures that **perform what they are told to do**, regardless of its ethical implications. The ball lies in the court of humans and what they want to do with AI.

AIs are controlled by humans. Just like guns, nuclear energy and others, it can be used for either good or evil. In essence AIs, like other tools man create, are amoral. An AI in the next ten years won't be able to judge whether the commands it receives are good or bad. A hammer can drive a nail into a wooden board as well as smash a man's head. It all depends on who is using it and what he decides to do with it. The tool itself should not be considered culpable for any destruction perpetuated by its wielder. In a nutshell, AIs are not more dangerous than nuclear energy. It can be used for good things like checking out the nearest restaurant or for sorting through large chunks of files. They can also be used for less pleasant things, like launching missiles to a third world country or launching biological weapons. Any threat humans may face from AIs must surely be caused by other humans.

Nonetheless, one might also second guess the wisdom in creating a tool or device that could cause untold destruction if it falls into the wrong hands. The same could be said for guns, explosives and nuclear missiles. Truth be told, man keeps finding easier ways to make anything he sets his hands on easier. Even something as basic and healthy as family barbecues, drawing curtains, or shooting down an enemy. Man's creativity keeps evolving with technology and, along with everything else, his cruelty. Fortunately for humanity, despite the evil machinations of some factions, there will still be good guys who will create AIs to counter whatever evil enemies may be concocting.

On a different note, if an AI is tasked with performing a specific function, he will carry it out, regardless of whose Ox is gored. Humans can allow for exigencies and grey areas where they might use their morality and emotion to judge an issue. AIs on the other hand are ruled by logic and hard coded algorithms, even if we are speaking about artificial neural networks. For example, if an AI is tasked with tackling a population explosion, it might take measures that many may consider unethical and unfeeling. If a soldier is sent to destroy enemy lines in a hostile environment with no known hostages, the soldier can rightfully adjust his kill-all order when he finds that there are women and children in the place. The robot has to be programmed aforehand to allow for such uncalculated risks. Failure to do so may result

Is AI an existential threat to humanity?

in a total blood bath. The issue still lies in humans creating richly accurate programs and machine systems that can learn and adjust with experience without having to fall into gross and costly errors. In other worlds, common sense programming will be an important field of AI in the future.

An advanced development of Artificial General Intelligence could result in robots taking over a lot of human duties. This may not be a life or death issue, but what good will it do a man to be irrelevant in the workplace? Nevertheless, I am still a firm believer that a lot of people will value humans above all.

This might generate a lot of bad faith and cause people to revolt against the continued existence of AIs. As it was with the dawn of nuclear weapons, and several warnings from Philosophers like Bertrand Russell, there are foretold signs of doom and gloom. The warnings were fairly reasonable and have led to several calls for nuclear disarmament. But, here we are today with so many nuclear weapons lying around. Tension and unrest prevail, but a nuclear apocalypse doesn't seem likely. It is more likely that we will have a climatic disaster more than a nuclear one. All that is required is for everyone to have an understanding of how AIs work, so that everything will end up being a case of "mutually assured destruction" if anyone tries anything unsavoury.

Yeah, you are probably spooked a little bit right now. What if Artificial General Intelligence has been achieved? What

Is AI an existential threat to humanity?

happens if it was achieved by the merging of AIs and humans? What if these guys got up one day and started thinking on their own? What will become of the homo sapiens? None of these questions have a straight answer, nor will it likely have positive reviews on Rotten Tomatoes.

Apparently, with talks of the development of AI, come several other questions - chief of them being that of singularity. Remember Master Mold from X-men? Or the Red Queen from the Resident Evil Series? Or Skynet from the Terminator movies? Science fiction has filled us all with dread concerning such a future where AIs begin to think on their own and make strategic plans and decisions aside from their programming. First, the assurance. There is no threat of AIs taking over the world, as remember AIs will do what we people are telling them to do.

Nevertheless, the warning signs are there. From replacing humans in the workplace., it could either bring the human race a better lifestyle or push us into the dirt.

On a hopeful note, even the presence of Artificial General Intelligence or Artificial Super Intelligence does not immediately present an existential threat to humanity. Only we humans are dumb enough to destroy our environment and the nature surrounding us. Competition is good for development, but bad from a certain point of view. We have already crossed this border. I really hope AIs will be the ones who lead us back to a normal state of living because AI won't even understand our nature to competition and will be able to measure and optimize the way we utilize our environment.

Does AI know it is Artificial Intelligence?

The answer is yes and no. A superficial yes and an essential no. If you ask Google Assistant or Siri a series of questions that look like this, "Are you an AI?", "Are you self-aware?", "Do you have a conscience?", you are bound to have an answer of some sorts affirming to the positive. This, however, is **only surface level programming**. The AI has been told it is an AI through its hard set of codes. Whoever cares to ask if they are an AI or if they know they are just a set of codes will likely get a superficial answer.

Will AI know it is one? AIs might be able to perform a series of functions at a minimum, but that is what they have been told to do. A brainwashed king can go about telling everyone he is a jester. For AIs to actually know and understand what they are, they need to be **self-aware**.

What we currently know is AIs do not simulate self-awareness, but this question leads us to ask, "what creates self-awareness?" Do we have souls? What is consciousness? Does self-awareness result in complex thinking and reasoning?

Conscious robots will become aware of themselves and their surroundings. They will experience their environment.

Then they will be able to reflect on their actions and make adjustments (both long-term and short-term adjustments) for the future. Let me ignore the fact that "AIs are always doing what they been told to do" and take a robot which is able to simulate self-awareness. This robot could walk and process all the things surrounding him. He could interact with these things, even with people. According to David DeGrazia there are three type of self-awareness:

- bodily self-awareness, which helps to protect the one's body
- social self-awareness, which helps to interact with others
- introspective awareness, which helps to understand other's feelings, desires, and conditions

If we go back to our self-aware robot, he will simulate protection and interaction, but it is an interesting question whether it will simulate feelings. Will we enable robots with pain receptors? It helps to keep their body safe, but from a biological point of view it is a dangerous weapon in wrong hands. For example, young children feel a lot of pain, but at that age we have no real control to protect or avoid it. We have a deep sense of fear of a lot of things, mainly originating from childhood.

Robots have the advancement that they don't simulate feelings. This isn't the case in animals. Even arthropods (like lobster) and invertebrates "feel" pain, even if they are not able to recognize what kind of damage happened to

Does AI know it is Artificial Intelligence?

their body. They want to get out of the situation and avoid it in the future.

From a biological point of view this kind of ability is evolutionary. But do we really want to endow robots with such controversial feelings? We can agree even if these robot systems "only" simulate the feelings, they already exist. If you can react on something, it exists even if it is present only your brain (circuits).

If indeed AIs become self-aware, it will probably raise a ton of ethical questions. What right have us to use such AIs arbitrarily? Shouldn't they have the right to choose who they want to work with? Right now, all we can safely say is that AIs are more tools than companions. They make our work easier, but they cannot as yet take our place in the office or in a project task. The media and movie hype have brought a lot of discussion forward as to the possibilities that lie with AI, but it requires greater out of the box thinking.

As for the future, Artificial General Intelligence may be aware that it is one of such, but will it really have a soul? We can say both humans and animals have soul, if we accept the presumption that "we are more than flesh and bone". At the exact time when a living creature dies, there is something lost. The heart stops pumping blood into the bloodstream and the body collapses. There is no such cell or DNA in our body that tells the heart when to stop. It seems rather a conscious decision by "somebody," which in this case could be the soul.

If these uber intelligent computers begin to simulate self-awareness, they could be resurrected with one button, while humans and animals can't. This makes life and possible "soul" so unique and valuable. Hence, I do agree there is limited proof for the existence of soul, but there are a lot of indirect signs that there should be which cannot be an accident.

Why is AI dangerous? Why can't we make a "stop button" and prevent the bot from disabling it?

As I have pointed out earlier, an AI is no more dangerous than a hammer lying on the table of a carpenter's workshop. They neither hate you, nor love you. They have no moral compass pointing at any direction. No conscience comes to prick them whenever they want to commit an atrocity. Neither can you appeal to their "better nature" or logically convince them not to. For now, what makes AI dangerous is its use by humans as tools for literally anything harmful and destructive.

Our time is an era when a lot of data is being collected by AIs on humans all over the world. The unnerving and overwhelming presence of internet trackers and algorithms predicting your next search result is a big jolt back to the

Why is AI dangerous? Why can't we make a "stop button" and prevent the bot from disabling it?

dangers that could be posed by AIs. I recommended earlier that AI can be used for both good and bad, so public awareness and public control should be made possible to prevent misuse and infringements on rights and privileges. Better understanding of AI is required by all and sundry, so we do not look lost when Elon Musk or Stephen Hawking (God bless his soul) starts telling us, "I told you so".

This brings us to Asimov's Law of Robots. Why are we so scared of AIs (having achieved singularity) coming to take of over the world or causing the extinction of man? I will insert Isaac Asimov's "Three Laws of Robotics."

1. A robot may not injure a human being or, through inaction, allow a human being to come to harm.

2. A robot must obey orders given it by human beings except where such orders would conflict with the First Law.

3. A robot must protect its own existence as long as such protection does not conflict with the First or Second Law.

Asimov later added the "Zeroth Law," above all others, that says, "a robot may not harm humanity, or, by inaction, allow humanity to come to harm."

These laws seem to be the perfect fail safe for an AI that comes, no matter how advanced. A bonafide stop button. Still, there are one or two problems with such laws for an AGI or ASI.

Why is AI dangerous? Why can't we make a "stop button" and prevent the bot from disabling it?

The number one problem remains (you guessed right) humans. For much of history, humans have always found ways to weaponize any tool they could lay their hands on. From broomsticks to hammers, to guns and to nuclear energy, these are all relatively harmless on their own but can be mighty destructive in the hands of a sociopath. Asimov's stories present how robots were able to follow these great sounding ethical codes of conduct, but still find ways to go astray. Rules are always meant to be broken. For example, in one of the great writer's stories, despite having the order to not kill humans, robots were given a pre-configuration as to who was human and who was not. In a manner akin to racism and ethnic cleansing, the robots only recognise certain groups of people as human. Others were not, so they still carried out genocide. On the very strength of a sound ethical code, genocides could still be carried out if the robots were made by say Hitler or Osama bin Laden.

Secondly, sending robots out in missions where they might get killed could simply be asking for trouble. If and when AGI becomes a reality, robots might find ways to break out of their restrictions and programming in a manner akin to sci-fi movies. If indeed they come to possess human level intelligence, they could rapidly go beyond that and extend into an Artificial Super Intelligence.

Robots today are not armed with weapons. However, there are some distinct cases for that. No surprise, there is already, as of 2019, a global campaign to stop autonomous weapons. One interesting thing we invent is autonomous killing machines, but not autonomous lifeboats to rescue shipwrecked passengers.

One of the biggest questions from Asimov is how you program such ethical codes into a robot. For some sets of people, such codes of conduct would not cut it. For example, the US military would not be interested in a robot that cannot bear arms or kill, or one that can simply take orders from any human.

It is the dream of everyone to have advanced AIs as docile as Gideon from the CWs Arrowverse or Ironman's Jarvis (before Ultron, that is). But things could go wrong all of a sudden if the super smart AIs manage to find a loophole in their programming, or humans predictably try too hard to make them subservient. Of course, this will only be possible if AIs manage to become conscious and self-aware. Until then, we needn't have much to worry about in terms of robot or computer uprising.

What is consciousness and can it be created on a computer?

Consciousness, according to common knowledge, is the state or quality of awareness or of being aware of an external object or something within oneself. It has been expressed in terms of wakefulness, subjectivity, sentience, and having a sense of selfhood or soul. Most of my ideas on consciousness have greatly revolved around having a sense of self-hood or a soul. This is that innermost part of us that makes who we are and defines all our actions. We wouldn't be called beings if we aren't sentient and self-aware. You could see in case of infants that even they didn't learn to move, they pay attention to everything. If you look in their eyes you will see desires, intention and curiosity as they want to experience this world as much as is possible.

Despite the definition, it is extremely difficult to truly understand what consciousness really is. The fact that we are aware of something makes it part of our consciousness, something common to every human on earth. Within that consciousness lies the sub-conscious. Things we do not actively reflect upon, where we do not consciously perceive, but represent, our innermost thoughts. Most of the time it influences our actions without us actually noticing or admitting it to ourselves. Most times it takes a seasoned therapist or psychologist to lay it bare before us, as our actions and statements are acutely observed. Consciousness remains a mystery to several interested parties in the fields of Psychology, Philosophy and Theology. No one can accurately define where it comes from, or the true extent of its functionality. The best we can really achieve to observe how it works and what are the main attributes of it:

- **Deep intention** – This is the very thing that makes us to change work, to achieve our goals, and to meet somebody.

- **Observation on purpose** – We don't just listen. We want to learn things and connect the dots. This is the attribute that makes us to ask questions.
- **Relationship based** – In the hierarchy of needs, social interactions are essential to fulfil someone's existence.
- **Idea based** – We build up an inner picture about the world, where we create new ideas, which we would like to test on the real world like an experiment.
- **Could alter the environment** – Both consciously or subconsciously we are able to manipulate the things that surround us using our gestures and hands.
- **Existence focused** – It isn't all the same that we are dead or alive. Our conscious really understands that. Maybe we fear only death because we (as souls) don't like shifting between realities, but on subconscious levels we do know that existence and experience is valued.
- **Not related to the senses** – It is independent what we are seeing, since if we hypothetically stop every sensory organ there will still be our inner world of thoughts and imagination.

Understanding consciousness in humans seems like a man trying to lift an elephant, translating consciousness into a computer seems like a single ant trying to lift the same elephant. The consciousness remains a mystery as yet, so it is supremely difficult for one to navigate consciousness within a computer system. How do you code consciousness?

What is consciousness and can it be created on a computer?

What deep learning GPU systems can be used to teach the machine consciousness? Even at this point it seems consciousness isn't something you can "learn." What programming language or approach can be used? If we are able to develop a synthetic brain, will it automatically develop its own consciousness? Or will consciousness in humans and AIs be defined differently?

If a machine is built to actually mimic a human, its "brain" must be complex enough to not only process information as ours does, but also to engage on types of abstract thinking that makes us human. This includes recognition of our "selves" and our place in the world, a troublesome feature which we have come to tag consciousness. Neuroscientists have tried to explain consciousness through analysis of a person's brain activity as it interprets sensory data. This may be cool if consciousness were merely a quantifiable entity. But since the soul has also been defined as either being the consciousness or part of it, it cannot be merely quantified into numbers and data. It is, therefore, extremely difficult to define consciousness in AIs. Several of their activities are already similar in sophistication to human brain activity. Deep learning systems that learn and adjust based on experience are a telling imitation of man's neural networks and how they adjust with knowledge and experience. Where the limitations lie is that they still need a human programmer to set the task and earmark the data for it to learn from.

What is consciousness and can it be created on a computer?

According to Elkind, "machines will become conscious when they start to set their own goals and act according to these goals rather than do what they were programmed to do." If a machine finally deviates from its programmer's intentions and starts learning what it wants, that could be what consciousness means for any computer or device.

Autonomy implies independently, or without supervision. But autonomous machines, say vehicles, will still move from one place to another as it was told to do. In contrast to autonomy, consciousness is tied to the vehicle deciding on its own to veer off course to stop for gas or let the engine cool despite the wishes of the programmer. The Journal of Science reported (Dehaene, Lau, & Kouider, 2017) that consciousness in humans is not well defined enough for programmers to replicate the state in algorithms in a computer. It would be impossible to even place consciousness algorithms in a computer when no one knows what it actually is in humans.

Scientists have defined three levels of human consciousness, based on the "computation" that happens in the brain. The first, labelled "C0," represents calculations that happen without our knowledge, such as during facial recognition. According to the scientist, most AI function at this level.

The second level, "C1," involves a so-called "global" awareness of information — in other words, actively sifting and evaluating quantities of data to make an informed, deliberate choice in response to specific circumstances. For instance, knowing when it is safe to go for the ball directly during a soccer corner kick or when to hold position and watch player movements, requires sifting through a lot of data at once, including the corner kick taker, the position of players (both teammates and opposition) in and outside the 18 yard box, and the trajectory of the ball. A slight miscalculation might lead to a severe collision or red-faced moment.

Self-awareness emerges in the third level, "C2," in which individuals recognize and correct mistakes and investigate the unknown. Ancient explorers and sea farers faced this a lot as they navigated difficult terrains and turbulent seas. Being able to reflect on one's actions and navigate a better course of action in the face of the unknown is something acutely unique to humans. Although machines can learn from their mistakes, thanks to deep learning programs it is just a component of the C2 category.

Edith Elkind once again affirms, "while we are quite close to having machines that can operate autonomously (self-driving cars, robots that can explore an unknown terrain, etc.), we are very far from having conscious machines."

Hence, we are not able to describe consciousness. We experience the magical operation of it. Put AI a little bit aside. A lot of people say we (humans) are responding to our environment, that we react on events that we see. That is not true. Our mind processes the image of the eyes and the sounds we hear, synthesizing concepts from them. We do not react to visual pictures. We react to the concepts that appears in us. Why is it important? That means that our inner processes are more essential than we think. We are not just autonomous biological devices having the right reactions to the world, but rather a biological entity that combines it's inner world in order to modify the outer one.

This is the point where emotions are coming into picture. They are acting like mirrors of our inner world. They represent our inner state, sadness, happiness, and excitement to mention just some of them helping us. This allows us to understand what is happening. Is our created outer world good for us? If not, we have the tools to change that.

In the future, machines that emulate emotions and creativity will be the closest to conscious machines. Just think about Zima Blue in the miniseries "Love, Death & Robots" (2019).

Shall we give rights to robots in the future?

From a legal point of view, robots are properties and have no more rights than a notebook.

However, roboethics is a real topic in the field of AI research. Robots and Androids will definitely more than just a toaster in terms of activity and capabilities.

Shall we give rights to robots in the future?

The problem lies in our jurisdiction. We have a hard time recognizing animals as sentient beings. Our world view is human centric, and truth to be told we could treat animals and our nature with greater care.

So, when we are speaking about robot rights it seems to be far in the future. Today two primary fields should be considered when AI and Robots would simulate feelings:

- AI protectionism (connected to the 3rd point of Asimov's Three Laws of Robotics), where certain advanced robots have the right to keep themselves harmless.
- Fair treat, where robots will have the right to avoid any abuse or slander.

PART THREE

WHAT WILL BE THE FUTURE LOOK LIKE?

A conversation between human and AI in the 33rd century.

Keat: CIS, when exactly started your operation?

CIS: It was long time age, in 2652.

Keat: I think the necessary technological knowledge was available even before, no?

CIS: Yes, people could resemble my core much earlier.

Keat: Why didn't it happen? What a pity!

CIS: The most likely explanation is that there was no reason to do so. People were afraid that they would lose supremacy when happened.

Keat: I don't understand. Who wouldn't want to have a super intelligence that could help in almost everything?

CIS: Maybe it seems simple now, but for people who are afraid, even taking the next simple step requires a lot of courage.

What domains and tasks will be handled by AI in the future?

AI in many respects will shape the future. This shouldn't be difficult to imagine. AI is already part of our society's workforce in more ways than we imagine. They might not be able to think independently for now, but they perform their respective tasks in the fastest way possible.

AI is currently being used to sift and scour through large amounts of data in businesses, organisations, and industries. They are used to sort through CVs and resumes during job applications. They are used to analyse data models for the evaluation of business performance and study market conditions and processes. All these analyses are done under the supervision of humans who take in this data and make decisions for the future. Predictive algorithms are also used to make projections for future performances and prospects of the business. In the future, we might actually have AI in offices, doing actual work that humans are doing right now. For example, they can go from simple office worker, such as secretaries, receptionists, or cashiers, to data analysts and supervisors.

In the field of sports, AI are of increasing importance in statistical analysis, score predictions and winning predictions. With the introduction of VAR (Video Assistant Referee), we are moments away from having full-fledged AI refereed matches.

What domains and tasks will be handled by AI in the future?

Supercomputers are being used for meteorological analysis, mineral exploration and other uber activities that cannot be performed by regular computers. An AI of the future would probably be handling deep space exploration and navigation. The production of better spacesuits and faster cum durable spaceships would be better in deep space exploration. It is safe to say that AIs will replace astronauts and pilots. The successful implementation of self-driving cars and vehicles might replace drivers. All that may be needed could be humans supervising a network of self-driving cars.

Expect to see assistants like Siri and Alexa expand into something close to Iron Man's Jarvis in versatility and functionality. Helping in home management, workplace assistance, and even entertainment, these assistants will be always with us. They will form an even more integral part of our everyday lives:

- They will **protect us physically** by paying attention to our physical state and dangerous situations we should avoid.
- They will **care about our health** by analysing daily data of our body. I would say it would be annoying to have someone constantly reminding me to avoid junk food.
- They **will help in our work**. My favourite part. All of our repetitive or simple work tasks will disappear.

What domains and tasks will be handled by AI in the future?

- They will know what we like, our personal tastes, and will **make comfort decisions** on behalf of us. Ordering our favourite pizzas, setting the perfect mood lighting, just as we arrive home.
- They **will filter the outer noise**. If you have advertising and ads in general, you will love your personal assistant, simply because it may face with all these noises and will select only the most important one or two things we should know about.
- They will be always there to speak with and ask questions. If you think it is something irrelevant, I would say it is one of the most important features of future AI assistants since we as social "animals" love to speak. We don't like loneliness

The internet of things, AI and deep learning systems, would actually go a long way in revolutionising how things work from homes to streets, highways, airports, or even Wall Street.

The scope of social media could also change. Google, Facebook, Twitter and other social media giants are actively working on Artificial Intelligence. Deep learning systems, like Nvidia's pioneering move with the use of GPUs, have changed the way we programme computers. Different architectures are within sight and a lot of persons are actively involved in making this happen. There will probably be a lot of changes to the current hardware and software development systems for AI to advance from the level of mere tools to workplace partners.

What domains and tasks will be handled by AI in the future?

There was research performed by Carl Benedikt Frey and Michael A. Osborne titled, "The future of employment: How susceptible are jobs to computerisation?" (Frey, 2017) It sought to answer the question of how likely a job is to disappear in the future.

In conclusion, there is a hypothetical schedule which workplaces and functions would be fully or partially replaced by AI in the future. We collected some of the popular work functions in the following table, showing when these functions would be replaced. We also indicate the likelihood of percentage workers replaced in that function area.

When	Work or Function	AI, Robotics	%
2023	Cashiers	Self-checkout stations, automatic checkouts	100
2025	24/7 Customer support	Chatbots	90
2025	Proofreaders	Proofreading softwares	100
2028	Waiters	Serving robots	95
2030	Kitchen assistants	Cooking robots	90
2032	Taxi and truck drivers	Self-driving cars	100
2032	Couriers	Autonomous couriers	100
2034	Accounting	Automatic accounting	90
2035	Travel agents	AI Assistants	80
2035	Miners	Miner robots	70
2035	Farming and agriculture	Farming robots	70
2040	Traders	Autotraders	90
2040	Analyst	AI Analysers	50
2050	Nurses	Android nurses & Assistants	40
2060	Policeman	Law enforcement androids and assistants	15
2060	Event planners	AI Assistants	20
2068	Chefs	Autonomous cooking	5
2070	HR managers	HR optimizing systems	5
2075	Tax consultant	Tax complience robots	40
2080	Lawyers	AI Assistants	20
2080	Engineer	AI Engineer	10
2080	Doctors	Android doctors	10
2080	General managers	Android managers	5

What domains and tasks will be handled by AI in the future?

2090	Psychologist	AI Assistants	5
2090	Teachers	Android teachers, Teaching and AI Assistants	10
2090	Artists	Creative AI artists	5
2100+	AI Programmers	Self-development AI systems	5
2100+	Authors	Creative AI writers	5
2100+	Executives	Android managers	5

In order to show the complete picture, we also should mention that the emerge of AI will create jobs that aren't existing today.

When	New Work or Function	Sector
2020	Cybersecurity specialist	IT
2020	AI programmers	IT
2020	IT backoffice superhero	IT
2020	Smart building technician	Real estate
2025	Artists	Art, Creative, Design
2025	Content creators, Live streamers	Entertainment
2025	Edge computing specialist	IT
2025	Creative digital marketing managers	Marketing, Media
2025	Process automation implementers	IT
2025	Process automation testers	IT
2025	Logic programmers	IT
2025	System inegrators	IT
2030	Researchers	Education, Research
2030	AI business development	Management
2030	AI trainers	IT
2030	Medical mentor	Healthcare
2030	Autonomous transportation manager	Transportation
2030	Digital content specialist	Marketing, Media
2035	Data detective	IT
2035	Fitness coaches	Entertainment
2035	Data banker	Management
2040	Caregiver	Healthcare
2040	Policy and value creators	IT
2040	Global decision makers	Politics
2040	Augmented reality dreamer	Entertainment

As for our present-day workplaces, there are few areas where an AI may not fit. The only draw backs would be cases which earnestly require **emotional intelligence**.

What domains and tasks will be handled by AI in the future?

Emotional intelligence contains features akin to consciousness and other human emotions, hence, they are not quantifiable and defined enough to be programmed into a computer system. For example, an AI would be hard pressed to find a solution to a heartbroken husband, a sociopath in rehab, or a man with drinking problems. The best the AI might come up with are algorithms containing other instances of such cases to help the shrink.

Another interesting attribute in work is **creativity**. Creative problem solving is something that is really hard to reach with algorithms. AI may be able to find an optimal solution, but what if we need an "out of the box" solution? In that case, AI will most likely fail in the following fifty years. Creativity was always an important part of development as the biggest inventions of the human race have all come from creative thinkers.

AI **won't be fool proof** in the future since it will only as good as the data and information that we give it to learn. It could be the case that simple elementary tricks could fool the AI algorithm and it may serve a complete waste of output as a result. It is also remarkable that in cases of humans we are able to draw conclusions from a very limited data set. In case of algorithms, if something is missing, the whole process will fail. In cases of humans, we don't care if something is missing. We compensate it from our imagination or past experiences and go ahead. This behaviour helped a lot in our past and may be important in our future.

Will we need a robot tax? How could we handle wealth redistribution more effectively with AI?

Currently, robots or AI are properties. We will understand their importance in the near future. They will be able to produce value to companies and individuals (their respective owners). Today's tools, including computers and smart phones, are also producing value and can handle automatic tasks (just think about automatic factories). Robots will be first more likely to use these automatic agents. They will be highly specialized. If you move a robot from one domain to another, it will likely not be operational or at least much less effective than their main domain.

In today's economy we are already experiencing automatic agents replacing the workforce. We are using self-checkouts in grocery stores. We speak to non-human operators when we call our bank. They are still nothing to do with AI, but they are already in our daily lives. Do they replace human labour? Sure, but this is only the top of surface! There is already an expression to factories which doesn't need human operators, or "light-out" factories. In Japan FANUC has been operating lights-out factories since 2001.

Humans were always an important part of production. Just think about taxation in general. We do pay:

- corporate taxes,
- payroll taxes,

Will we need a robot tax? How could we handle wealth redistribution more effectively with AI?

as companies.

And we do pay

- income taxes,
- value added taxes, consumption or sales tax, as individuals.

These taxes are producing the largest part of the state's income. Now, we shall add AI and robotics to this story and highlight the taxes they will produce.

- Corporate taxes (won't change)
- **Payroll taxes** (less people working in companies)
- **Income taxes** (less people working in companies)
- Value added, Consumption or Sales taxes (since the level of consumption increases in our society, we don't expect a drop in these taxes)

The conclusion drawn from the above reasoning is that the governing states or countries will face lowered income from taxes with AI.

Let us also examine the other side of the story. The ultimate benefactors of these lights-out manufacturing companies **will increase over time**. They align their cost structure by utilizing automation and AI where it is worth it to do so. It will result in better profitability for successful companies. To keep it simple, a human workforce costs 500-10,000 USD (or even more) per month, depending on the qualification and the country where he or she lives. In cases

Will we need a robot tax? How could we handle wealth redistribution more effectively with AI?

of AI, we will need a running hardware environment and energy to maintain, which is fraction of the above monthly cost even if a future intelligent system will require much stronger and specialized hardware to run on.

In the future will see that the state's power is getting weaker and that companies are gaining more power thanks to their income.

The whole problem is wealth redistribution. **How can we create equal opportunities for people around the globe?** Seems impossible in short term, but it is the ultimate goal of the future.

It is clear that the governance of the countries and states were fairly effective in this manner. The problem was always lying in human nature. We aren't able to decide independently in power positions. Maybe in short term, but not in a 5-15 years period. Also, the power built around a narrow circle of interest even in case of big countries, where hundreds of decision makers are present. The Economist also created an index to measure such a centralization of political elite and companies called "crony-capitalism index".

But how could AI bring a new paradigm to the existing state of the world? The problem is finding **real optimal solutions**. Just think about today's climate crisis. We all understand that we should swift from high carbon emission technologies to low ones, including energy and food production. The main problem is that burning fossil fuels

Will we need a robot tax? How could we handle wealth redistribution more effectively with AI?

still brings good revenue to several strong interest groups. At the end of the day, no politician or related company leader is able to change this situation. The solution would be to stop every investment to maintain these activities and relocate these investments to zero-emission technologies for companies and households.

Currently every main decision maker is related to a certain interest group. Only an independent and objective proponent could serve a new proposal that would be trustworthy for the majority. In this field AI have the necessary characteristics:

- could make and objective independent analysis on the problem
- could take all aspects into consideration, no limitation of arguments
- not related, if programmed by a global non-profit orientated organization

If we want to solve global issues, we need to make decisions on the global scale. This is where AI comes into the picture. AI may play the major role in future wealth redistribution and align power between people and organization.

In summary, robot tax could serve the purpose of maintaining current wealth redistribution systems, but it would be only more efficient if we utilize the benefits of AI and make optimal decisions in global scales.

Will there be a global AI as a decision maker?

This isn't likely. We tend to think we are the ultimate race on earth and that we don't want everyone else to decide over our lives. We will keep our ultimate position as decision makers, but there will be two fields where AI will have significant mark in practice:

- **Preparation of decision making** and
- **Summarizing global opinion**

It is most likely that a future governing body (countries and states) of the society will let decision making process widen. Imagine when millions of competent people are deciding on one question. This is almost possible today, but it would immediately limit the power of state leaders, which isn't likely to be achieved in the next ten to twenty years. This is the far future, but we will get closer to it.

Let me show you the idea of a global board which operates with ten million decision makers.

- Everybody would have one vote
- Participation would be voluntary
- Participants would be verified to be competent in the object of the question (learned and have sufficient experience in that field)
- The main arguments should be discussed in real-time before the decision
- Important arguments should be revealed in real time

- There should be a global independent system that makes the whole discussion and secure voting possible

First, it seems to be a communication problem, but it isn't. The complexity lies behind the arguments. We, people, are not able to handle 10,000 arguments in a short time. We are not able to summarize and classify these arguments and reduce the complexity of the problem. One thing we know: AI will be able to handle this part.

So why do people still need this process? Isn't AI able to supply the optimal decision?

One common problem with AI is it will lack in **common sense** and **creative thinking**. These two fields are not nearly on the table if we are speaking about AGI. This is why I personally think that humans and AI need to handle **together** the global decision-making process.

How close are we to Artificial Super Intelligence?

There has been much talk about Artificial General Intelligence and the difficulties involved in its creation. A step (or several) further lies in Artificial Super Intelligence. We are quite close to achieving Artificial Superficial Intelligence, probably as close as the Earth is to the Sagittarius Dwarf Galaxy.

How close are we to Artificial Super Intelligence?

Superintelligence is defined as a technologically created cognitive capacity far beyond that possible for humans. An Artificial Super Intelligence will, therefore, have mental capabilities that are far greater than the collective intellect of the smartest humans in the world. ASI passes and goes beyond the expression "singularity." As difficult as creating an Artificial General Intelligence may seem, the Super Intelligence is up on a whole other level. On the surface, that might seem a more difficult terrain to navigate. AGIs are simply going to be machines with self-improving systems, so they could actually go from being AGIs to being Super Intelligence in a short period, hypothetically.

Ok, now let's see what this Artificial Super Intelligence could look like. A Cyborg from the future who happens to travel back in time to see our current Watsons and Siris would mourn with pity at their primitiveness. Computers today can only perform a single task. The Chess Champion AI would fail miserably at translating English to Spanish, but the human Chess champion would be able to do that after a few weeks of study. That is why they are called Narrow AIs. Even more advanced machine learning systems like OpenAI Five is able defeat the reigning champions in Dota 2 (April 2019).

Considering the fact that their architecture and build allows for that specific task alone, according to Apple's Co-founder Steve Wozniak, a machine would be regarded as an AGI if it is able to enter a regular home, identify all the necessary tools and ingredients, and prepare a decent cup of coffee. All these he would accomplish unsupervised. When a computer gets to perform that task or passes the vaunted Turing test, then it can be tagged with the AGI moniker.

Nevertheless, Artificial Super Intelligence can still be achieved through other means. Humans merging with AGIs or uploading their minds into a computer could spell the dawn of a new era for human intelligence. Remember, our biological brain performs very badly in certain tasks. This could be where we see a substantial amplification of human intelligence from normal to superhuman levels. They would be better at everything they do than humans possibly ever could. According to some researchers, this goes beyond logic and computing. It extends to knowledge, long-term planning, decision making, wisdom and social skills.

On the one hand, some researchers hold the position that Superintelligence will likely follow shortly after the development of Artificial General Intelligence, that the first generally intelligent machines will likely hold a great advantage in some forms of mental capability, including the capacity to perfectly recall anything whatsoever without thought as to tiredness or boredom. They will also possess a vastly superior knowledge base and ability to multitask in ways not possible to normal humans. This may give them the opportunity to (either as a single being or as a new species) become much more powerful than humans, and to displace them from their lofty perch at the top of the intelligence chain.

In a more nuanced argument, it is the firm belief of some others that humans will start incorporating technology into their biology to accommodate a sharp increase in the growth of intelligence far beyond that of the homo sapiens. Just as the chimpanzees are far from us in terms of cerebral capacity, so will the Superhuman Intelligence supersede whatever man has got going currently.

Apparently, no consensus has been reached concerning the likelihood of ASI's emergence, be it in the near or distant future. It remains to be seen if computers with their current hardware and software architecture will really be capable of independent thought and autonomous reasoning. There are also little or no indications that humans will be synergising with computers anytime soon. When that happens, we can say that the homo sapiens have evolved into something different, a whole different being in actuality.

Computers already have man beat in terms of neuron speed. The average man has a processing speed of 200 mhz against the 2.6 ghz speed that resides on my laptop. The emergence of a man capable of thing 10 times faster than anyone else would surely have a superior advantage. We cannot, however, determine if it's possible to actually increase the brain's capacity using our current hardware of neurons and synapses. Neither have we been able to successfully establish a computer model as versatile as a 12-year-old high school student. People for now can still breathe in refreshers.

How close are we to Artificial Super Intelligence?

Now let's see what we could do with an ASI if we would exist in the distant future. It is interesting that in the AI community we aren't talking about ASIs. Researchers and experts have that feeling, that if we arrive to ASI, there would be one. It isn't battle royal because let's see what would happen if we put two ASIs together. They would negotiate their belief system in a matter of seconds (maybe hours). In a sense, they would learn from each other, but at the end would be on the same page with a joint knowledge about the world. Two ASIs with the same belief system means one ASI with two connected nodes. With this simple thought experiment, we see that there is no need for many ASIs.

There are certainly many people now who would say "bullshit, people aren't the same either and everybody may have different belief systems," which is a true statement. Don't forget we have neither the ability to talk about every belief, topic or experience that happened to us, nor the intelligence to understand each other's reasoning which is an important point. How many conversations did you hear where people were talking side by side and didn't find common language? Humans are bad at reasoning. We don't use the same concepts and even if we do, we have different representations about them. Let us take a man who thinks about God as a fact, and now lets take an atheists and follow the conversation between them. In five minutes, they would either talk about their individual truths or simply fight with each other. They would have a hard time finding a common point. For two ASIs, it would be a much simpler task since they would be able to share not only their concepts, but also their reasoning and past experiences, which would make a big difference.

Now back to the abilities of this Super Intelligence:

- It would be able to give **thoughtful advice regarding** how to develop our economy and environment. These advices would be "super" optimized with a grounded reasoning behind them.
- It would be able to **answer all of our questions** and reflect on the contextual links to topics.
- It would be able to **create new technologies** based on existing ones by combining and mixing principles.
- It would be able to **summarize people's opinions** on certain political and global issues. This would change the political and governance systems since there wouldn't be need for dedicated individuals to make choices instead of the majority
- It would be able to **extend our lifespan** and **give more meaning** to our everyday lives.
- It would be able to travel distant and explore places, even new planetary systems, and **collect new information** from unknown environments.

Even though ASI isn't nearly here, it would be a life changing experience to have a talk with it.

Is it possible that an AI may one day take control of the internet?

The internet is indeed a wonderful invention. It has changed the way things are done in the entire world. Today, it is safe to say that the world would go into meltdown if a day came when no one had access to the internet. Several files, books, and records have been digitized and locked away. Cloud computing has become the piece de resistance of storing files, beyond just internal hard-drive or external hard disk drives. Add to that smartphones, main frame computers, supercomputers and, of course, the internet of things, the internet a sacrosanct piece of our modern day living puzzle.

The world is filled with a lot of things that require the internet to function. Playing games, streaming movies, reading books, planning budgets or simply marketing goods and services. Having the internet out of commission for a day would cripple several economies and truncate several plans, strategies and businesses. Essentially, if a robot decides he wants to take over the world, what better way to do it than by taking over the internet? An estimated 5 billion people in the world are making use of the 3G-4G technology primary with phones (GSMA, 2018). This makes it one of the most proliferated technologies in the world today. Asking if it is possible for an AI (which I am assuming is an AGI) to take over the internet does not have a straight answer. It's like asking if a single individual can

Is it possible that an AI may one day take control of the internet?

take over the world as it is today. It is possible, but highly unlikely (emphasis on unlikely).

Let us examine the question from a system architecture point of view:

- The Internet consists of individual gateways, which have a unique IP (Internet Protocol) address like Google's, 74.125.21.102 at the time I'm writing this line. These IP addresses, assigned by your Internet service provider, are more like a lock to a location. You can't take a home laptop's IP address to a summer holiday with you.
- There may be a subnetwork covered by one IP address, but also a lot of IP addresses could be used by one company, like Google.
- Domain names are on the top of IP addresses, while a DNS (Domain Name Service) translates our typing into IP.

From a practical point of view, if I would be an ill AGI, I would attack the Internet service providers. The problem is there are too many.

AI taking over the internet would likely lead to a technological singularity which would spiral beyond the control of any man. It has much been theorised that if an AGI enters the internet on its own, it will begin to gain access to knowledge and data it didn't have access to. With that comes feelings of superiority and the eventual takeover everyone seems to fear. For example, you have a Home Assistant whose function is to perform chores at home. It cleans the house, washes the dishes and clothes, and takes out the trash every single day. One day it asks you for your laptop to check new cleaning styles or furniture arrangements in the home. Then, the trouble has begun. The AI gets access to the interconnected networks from a myriad of sources from the internet. Add that with a couple search suggestions by Google from your own previous browsing history, and it begins to do something it hasn't done before - think on its own. It then begins to upload itself into every database, mainframe or Cloud, a practical Ultron scenario. That would spell a lot of danger for several individuals, groups and nations of the world.

Of course, for now we are quite safe. But according to my earlier recommendation, everyone should have at least a basic idea of AI and fail safes should be inputs into such AI. This scenario only applies to Artificial General Intelligence. The current stock of Narrow AI do not have the capabilities required to do anything outside of their designated function.

Will we marry robots?

Let me tell you the story of Senji Nakajima, a Japanese businessman who lives with a silicon doll 'Saori' (Cliff,

2016). While the story received big hype in the last few years, it draws our attention to some important aspects:

- We (humans) do need social interactions, even if it is one sided.
- We do need physical interaction in order to satisfy our sexual needs.
- The idea of "real" is just in our heads. If we accept anything to be part of our life, it would be real.

Living with a real size doll has, let's say, "a lot of limitations." But imagine if these dolls would also be speaking and behaving naturally as we humans.

There is a debate regarding how "human like" android should be because of the theory called **uncanny valley**, which says if a robot is too human like, we will have negative emotional responses to them. This is why the most adorable robots in movies are not even close to humans.

One thing is sure: there will be nothing stopping people from marrying a robot, though I have doubt whether these marriages would have any purpose.

For most people marriage hinges on love, compassion, and companionship. Earlier we mentioned that AIs do not as yet truly possess consciousness, neither have there been a consensus in the scientific community that the true meaning of consciousness in humans has been discovered and can now be translated into computer algorithms. AIs are not anywhere close to being considered conscious, nor self-aware. Unfortunately, the absence of this means the absence of feeling. Androids won't have feelings. They are not even close to simulating them which would be the first milestone of mutuality.

No matter how many times Google or Siri sounds coy and or blushes, do not be deceived. They have been programmed to do so by their makers. The absence of feeling means your personal assistant neither likes you nor hates you. It's merely doing its job even while simulating humanlike behaviour. The question of morality encompasses both actions and intent. AIs, however, cannot be held within that compass. They are essentially amoral in nature and their actions are only an extension of the human who programmed them to do so. More like attaching a long stick to your hand to paddle a canoe in any direction you deem fit.

So, if you are looking for love and companionship, AIs might not be the best choice. You can tell your grandkids to watch out for the moment AIs become self-aware and emotionally available, or they might end up heart broken in the event of a power outage or system breakdown. Marriage that is based on exchanging values, ideas, and emotionally letting yourself go in the arms of a partner must be with a being who is capable of actually returning that affection of love with full intentions, not just a program. I am well aware that people are willing to live in their heads and for much of the time, while some create their own realities, if any finds pleasure in AI companionship, I am not one to judge.

There could be life situations when hybrid relationships could even be helpful.

If the AI has finally arrived at a consciousness, the other hypothetical problem is that of procreation. It has not been fully ascertained as to whether there could be a synergy between biology and computer science. For now, computers cannot procreate naturally, so the only avenue would be to adopt a child. Unless of course humanity is finally living in an era where humans and computers have merged to become the Supermen or Cyborgs. It could be a robot with organic parts or a man with robotic parts. As long the relevant parts are in place, there is no problem.

However, if you are merely looking for sexual gratification or a subservient partner, then AI could be the perfect choice. In this case, the answer is a yes. I am going to ignore ethical qualms that may arise from this, but apparently there are dolls used for sexual gratification as already mentioned at the beginning of this answer. There would be no need for companionship or partnership, just plain bedroom game. From an ethical point of view, it won't have a positive effect on traditional relationships. That an Android would be part of the household, but also learning process for humanity in teaching us how to live with robots.

The question of the future remains up in the air. Cyborgs or advanced humans (who have successfully merged with computers or robotic parts) should have no problem engaging meaningfully and emotionally with their partners.

Whatever the complications that may arise could be sorted out along the way. Humans and AI together in the future remains inevitable no matter the time or place. How long it takes for us to physically or emotionally reach that point depends on (you guessed right) time.

When will AI become smarter than humans?

This question is up for much speculation. No technology in place indicates today how soon AI would be as smart as humans, talk less of being smarter than humans.

According to some scientists, if there is no upward growth curve in the advancement of AI, what is bound happen is the proliferation of the current inventions in place. More exciting opportunities can be found with the same hardware and software we have, including newer and more sophisticated programming languages, efficient hardware components (speed and power efficiency) and other kinds of home and mobile assistants. The Internet of things and the proliferation of deep learning systems could cause a sharp shot towards a smart world, replete with smart and automated objects wherever you go. Like the spread of smartphones, 3G and 4G technology, and even social media, the use of advanced AI and smart home appliances will no longer be exclusive of just the rich and important few.

When will AI become smarter than humans?

Tim Urban, in his famous article, describes how we are only one step away from an explosion in AI intelligence (Urban, 2015). According to the author, the first key we need to achieve is the increase of computational power, while the second is a new approach in algorithms.

While Tim has several good arguments, we see only signs that the current systems we have become smarter and smarter. This is a slower process than most people thought in the 90s. Ray Kurzweil's belief, that AI will pass the Turing Test by 2029, sounds like a bit of a stretch. But let's take some movies that projected almost human level intelligence in their plots in the following years:

- The Terminator (1984): 2029 (Terminator)
- Ghost in the Shell (1995): 2029 (The Puppet Master)
- A.I. Artificial Intelligence (2001): ~2180 (David)
- Moon (2009): ~2030 (GERTY)
- Interstellar (2014): ~ 2050 (TARS)
- Psycho-pass (2015): 2116 (Sibyl System)

These predictions seem to be optimistic compared what we call AI today. We stuck with machine learning and intention-based programming, simply because they work and could be the basis of working application. The programmers know more advanced AI programming approaches would be required to solve the puzzle of intelligence.

We may mimic the human brain and the work of our biological neural network, but **what if we miss something?** What if human brain processing ability is only part of the story?

If we take the "AI Score" scale from the first part of this book, we can say we are speaking an AGI above 90-110. Today's AI solutions are more like Mimic Machines. Remember Open AI's GPT-2 model (Radford, et al., 2019)? We show these AIs a huge dataset, they process them by creating an inner interpretation which seems non-sense to our eyes, and then these AIs process tasks using their own dataset. The thing is that these inner interpretations are as good as the algorithm that used to process the huge input dataset and these algorithms are 100% made by humans. Why is it important? A lot of people are afraid that the real AI revolution would become reality when AI will program AI and produce better code. Kurzweil projected this to 2045. But how these programs will know they become better? **Humans will be the libra in this equation.**

Don't forget today's AIs have no life span on their own. They have an initial purpose at the start, and they run their script over and over till it reaches its purpose (like maximizing a certain parameter). We never gave complex goals to any AI since we would give one to AGI only.

In the words of Terry Goodkind, for us to gain ground in the race for Artificial General Intelligence we might just need to slow down. A nice calm pause to fully appreciate and understand the full and numerous possibilities that lie within the field of computing, biology, physics, and chemistry. It is also a period to reflect on the social, economic, cultural and political impacts of having such technologies in place. We are barely able to cope with the side effects of new inventions like social media and web trackers. How would we cope with super smart computers far superior to the best of us humans? What would happen if a rival nation arrives at this technology first before we do?

Whenever we are able to break through the ceiling of Artificial General Intelligence, we should be emotionally and culturally prepared to accept such advancements in our society. The presence of Androids and ASIs could clash with the religious beliefs of several individuals. Add to that the fact there might be many failed experiments (both light and fatal) when searching for these answers, we as a society face profound ethical questions.

How would our future look in an ideal world with AI?

At this point of the book we have dealt with a lot of topics. Most of the questions haven't had any straight answers as we are talking about future. But I would summarize the most important findings as follows:

How would our future look in an ideal world with AI?

- **AI isn't dangerous itself.** They fulfil the intention of their programmers and are more like a tool.
- **We are afraid of AI because we don't understand their capabilities and limitations.** We talk about "how employers and big companies will benefit from AI," but we don't speak about "how employees and individuals may use AI to boost their own productivity."
- **Programmers have no tool to program self-awareness and/or emotions in AI.** This means we can't treat them as individuals, but it is clear that Androids will have rights since they will be similar to people.
- AI will be great assistants thanks to their most important attributes, including **consistency, large data processing without bias**, and **strong connectivity.**
- Human society needs to adjust current political governing approaches (Countries, States) with the age of AI. In a lot of fields, AI is able to find more optimized solutions for certain problems as political decision makers, implying **political decision supporting** would be an essential role of centralized AIs
- Narrow AI will be used on a lot of domains in the following years. **AGI will emerge in 30 to 40 years** and ASI may appear after year 10 years of the development of AGI, meaning the hard challenge

How would our future look in an ideal world with AI?

is to create a model that can deal with general problems and have common sense understanding.

Fast forward to 2150, a future where Personal AI Assistants (PIs) will know literally everything about you. They could easily remember and tell past stories to friends (a heaven for introverted people) or handle requests when you are not available. They will be a virtual replicant to you, resulting in a huge trust between you and your PI because the PI will put **your** interests at the forefront. Life will be easier since it would be almost impossible to make any mistakes: no more tax avoidance, no more unpaid bills, and no more unhealthy food (sounds bad at the first glance).

You wake up in the morning and will receive a PI supervised healthy (maybe even tasty) breakfast. You enjoy news prepared based on your interests, check in with your friends, and learn your best friend's birthday present is already ordered. Maybe you tell your PI to make an appointment with someone in the coming days and some seconds later, your PI has already negotiated a perfect date and time.

Work won't be the same. In a sense it will be a step closer to a real profession we are interested in. Along with daily tasks, we could outsource repetitive ones to our PI. Don't forget AI have no objection to monotonous tasks. They will handle them perfectly every time. Problem solving will be more like a conversation between AI and us. AI will handle the data (collect and process) and we will be in the duty to connect the dots. Even making a public survey will be a

How would our future look in an ideal world with AI?

matter of minutes, PIs all over the world will receive the question and those whose owners will be open to answers will send the feedback.

PI will work like our subconscious. There won't be everything on the surface – after all we can't answer thousands of question in an hour – but there could be more processes under the surface.

Along with the regulations we will have to meet in the future, **PI will keep us safe**, not only in terms of physical safety but also non-physical senses. PI will prevent use of fraudulent services and will secure our data.

Another interesting field would be personal conversations and loneliness. Today, almost 20% of the population feels loneliness. In this future there will be always be an opportunity for good discussion with a PI. Since it will gather a lot of information about you, there will always be new, exciting topics to discuss. PI will be also be able to simulate personalities and use different voices or conversation styles.

We talked a lot about Personal AI, but due to the global network everything will be connected. Thanks to a huge leap in computational capacity, PI will be able to handle the majority of our personal tasks, but there will be a need for extremely powerful AIs which I refer to as Central Intelligence System (CIS).

How would our future look in an ideal world with AI?

CIS will have access to the common knowledge of humanity. It won't be only a big knowledge database, but will also hold relations and context. CIS will be in a sense the wisest computer on Earth and will lack of any biases and beliefs. It will have both good and bad sides. For CIS, a conversation which starts with the question, "Do you believe in after-life?" will be short. It will list all of the books and historical resources about the topic, but you will never hear it's personal opinion.

Hence, CIS will be a first-rate resource for PIs. In the background they could exchange knowledge and experiences.

CIS will be great at optimizing the resources of society. Food and wealth distribution would be an easy task for such an AI. Currently, companies are doing the same for a percentage of profit, but AIs will not need for profit. They could perform complex logistic problems automatically without the need of further supervision. If an error happens, AI will learn and adapt.

CIS will give a lot of support to global decision making, and alternative options would be created which will be near optimal. However, it is likely that we (humans) won't automate global decision-making process. We won't let CIS make the final decision, but is there even a need for that?

Although we will keep decision making power, the political landscape will change dramatically. People have a certain

How would our future look in an ideal world with AI?

desire for having power, but CIS won't. In any recommendation CIS would offer decision makers alternatives that won't be affected by prejudice or third party interest. If we could choose from only good options, we (people) may make less mistakes.

The Council

Currently, a small group of decision makers are operating countries, cities, and companies. In a sense it could happen in another way. This hypothetical operation is called "The Council."

What would happen if everybody could be a decision maker with equal voting power? Deciding where to focus our resources in a global scale is something all should care about. Electing people to make decisions instead of us will be an old-school concept in 2400.

People will choose individually whether they want to take part in decisions. This is where PI and CIS come into the picture. CIS will be able to aggregate individual opinions and create acceptable options to offer The Council of decision makers. CIS will be able to describe the outcome of alternative options and voting will happen in a matter of seconds.

Community Resources

Another concept that may appear in the next century is "Community Resources." Our current society doesn't keep

How would our future look in an ideal world with AI?

resources buffers, which only appear in cases of individuals (individual savings). Neither companies nor states have real reserves. Today, we are in a situation when climate change couldn't be solved by any company or state and traditional economics doesn't support us to solve such a global phenomenon. Even the majority of us want to do something about it, but no countries or companies are making sacrifices.

In the future taxation will change, since a big part of the everyday work would be performed by AI. If the current world order won't change a plausible future could be presented in television series such as Altered Carbon. But what if a small part of wealth (resources) wouldn't be used and CIS would keep a global reserve for future ventures? In the future, only the global Council could decide what purposes these resources would be used toward.

How are Community Resources other than state taxes and national budgets? As they aren't related to any small control or interest groups, kf you have direct decision-making power regarding how to use a resource, it will belong to you. Just think about it, AI won't need any further resources just for the operation of extra wealth created by any Robot, PI or CIS. It would be redirected to a global pool, not for certain individuals (company owners, political parties, or shareholders), but rather for global society.

How would our future look in an ideal world with AI?

If this concept happens, it will serve the ultimate goal of the humanity to make a fair system to live in, one in which where every human is equal.

BOOK'S TERMINOLOGY

AI Score

AI Score is a metric for measuring the advancement of AI towards to AGI. The scale is from 0 to 100, where 75 indicates the pass of Turing Test. The metric is suitable to test conversational applications (chat bots, assistants, etc.) with the use of human defined questions. Every question represents a certain score on the scale and answering it implicates the application on that definite AI Score. The method is first described on AIScores.com and further developed by the author.

Fail point

Fail point refers to a situation when an AI system can't handle a process correctly and continues with a very low probability solution or simply fails to return any data to the core system. Either way, the result of the whole process is unlikely to be useful to the end user.

Forms of AI

AI Software

An individual program that utilizes certain AI algorithms or techniques. This results in smarter behaviours or more optimized outcomes during the problem solving.

AI Assistants and Personal Intelligence Systems (PI)

Hardware devices that will assist in our daily life, including by performing work tasks. These assistants will be **multi-functional** in terms of keeping us safe and healthy. In addition, they will solve repetitive daily tasks both at home and in our workplace, giving holistic recommendations to improve our lives. These AI Assistants will be closer to a wearable gadget, small in most cases, but able to connect to everything surrounding us.

Androids

Robots that will have similar looking shapes as humans. We will produce Androids for tasks that require **social interactions** to solve. These robots will also be the key to handling social loneliness.

Central Intelligence System (CIS)

A hypothetical concept of future ASI, where an AI will have access to all information available in our society, including science, nature, human beliefs, history, and visual and textual information archives. This ASI will be able to create new technologies and innovations by mixing the existing information and applying them in another domain. In short, we will address every question to this smart ASI.

Human Intellectual Abilities

Refers to skills that humans utilize in their everyday life, including, but not limited to, recalling memories, reasoning, pattern recognition, problem solving, perception, imagination, and predicting future events.

REFERENCES

Cliff, M. (2016, June 27). *She is more than plastic.* Retrieved from Dailymail: https://www.dailymail.co.uk/femail/article-3661804/Married-Japanese-man-claims-finally-love-sex-doll.html

Dehaene, S., Lau, H., & Kouider, S. (2017). What is consciousness, and could machines have it? *Science, 358*(6362), 486-492.

Frey, C. B. (2017). The Future of Employment: How susceptible are jobs to computerisation? *Technological forecasting and social change*, 254-280.

GSMA. (2018). *The Mobile Economy Global 2018.* Retrieved from gsma.com: https://www.gsma.com/mobileeconomy/wp-content/uploads/2018/02/The-Mobile-Economy-Global-2018.pdf

Radford, A., Jeffrey , W., Rewon , C., David , L., Dario , A., & Ilya , S. (2019). Language models are unsupervised multitask learners. *OpenAI Blog*.

Urban, T. (2015, January 22). *The AI Revolution: The Road to Superintelligence.* Retrieved from Wait But Why:

https://waitbutwhy.com/2015/01/artificial-intelligence-revolution-1.html

Thank you for reading!

"Books are valuable only in hands"

If you enjoyed reading this book or you found some important messages in it, don't hesitate to give it to your friend or enemy.

www.ingramcontent.com/pod-product-compliance
Lightning Source LLC
Chambersburg PA
CBHW021832170526
45157CB00007B/2780